Springer Series in
OPTICAL SCIENCES

93

Springer

New York
Berlin
Heidelberg
Hong Kong
London
Milan   **Physics and Astronomy**   ONLINE LIBRARY
Paris
Tokyo   http://www.springer.de

Springer Series in
OPTICAL SCIENCES

The Springer Series in Optical Sciences, under the leadership of Editor-in-Chief *William T. Rhodes*, Georgia Institute of Technology, USA, and Georgia Tech Lorraine, France, provides an expanding selection of research monographs in all major areas of optics: lasers and quantum optics, ultrafast phenomena, optical spectroscopy techniques, optoelectronics, quantum information, information optics, applied laser technology, industrial applications, and other topics of contemporary interest.

With this broad coverage of topics, the series is of use to all research scientists and engineers who need up-to-date reference books.

The editors encourage prospective authors to correspond with them in advance of submitting a manuscript. Submission of manuscripts should be made to the Editor-in-Chief or one of the Editors. See also http://www.springer.de/phys/books/optical_science/

Editor-in-Chief

William T. Rhodes
Georgia Institute of Technology
School of Electrical and Computer Engineering
Atlanta, GA 30332-0250, USA
E-mail: bill.rhodes@ece.gatech.edu

Editorial Board

Toshimitsu Asakura
Hokkai-Gakuen University
Faculty of Engineering
1-1, Minami-26, Nishi 11, Chuo-ku
Sapporo, Hokkaido 064-0926, Japan
E-mail: asakura@eli.hokkai-s-u.ac.jp

Karl-Heinz Brenner
Chair of Optoelectronics
University of Mannheim
Institute of Computer Engineering
B6, 26
68131 Mannheim, Germany
E-mail: brenner@uni-mannheim.de

Theodor W. Hänsch
Max-Planck-Institut für Quantenoptik
Hans-Kopfermann-Strasse 1
85748 Garching, Germany
E-mail: t.w.haensch@physik.uni-muenchen.de

Takeshi Kamiya
Ministry of Education, Culture, Sports
Science and Technology
National Institution for Academic Degrees
3-29-1 Otsuka, Bunkyo-ku
Tokyo 112-0012, Japan
E-mail: kamiyatk@niad.ac.jp

Ferenc Krausz
Vienna University of Technology
Photonics Institute
Gusshausstrasse 27/387
1040 Wien, Austria
E-mail: ferenc.krausz@tuwien.ac.at
and
Max-Planck-Institut für Quantenoptik
Hans-Kopfermann-Strasse 1
85748 Garching, Germany

Bo Monemar
Department of Physics
and Measurement Technology
Materials Science Division
Linköping University
58183 Linköping, Sweden
E-mail: bom@ifm.liu.se

Herbert Venghaus
Heinrich-Hertz-Institut
für Nachrichtentechnik Berlin GmbH
Einsteinufer 37
10587 Berlin, Germany
E-mail: venghaus@hhi.de

Horst Weber
Technische Universität Berlin
Optisches Institut
Strasse des 17. Juni 135
10623 Berlin, Germany
E-mail: weber@physik.tu-berlin.de

Harald Weinfurter
Ludwig-Maximilians-Universität München
Sektion Physik
Schellingstrasse 4/III
80799 München, Germany
E-mail: harald.weinfurter@physik.uni-muenchen.de

Takahiro Numai

Fundamentals
of Semiconductor
Lasers

With 166 Figures

 Springer

PHYS

Professor Takahiro Numai
Department of Electrical
    and Electronic Engineering
Ritsumeikan University
1-1-1 Noji-Higashi, Kusatsu
Shiga 525-8577
Japan
numai@se.ritsumei.ac.jp

Library of Congress Cataloging-in-Publication Data
Numai, Takahiro.
    Fundamentals of semiconductor lasers / Takahiro Numai.
        p. cm. – (Springer series in optical sciences ; v. 93)
    Includes bibliographical references and index.
    ISBN 0-387-40836-3 (alk. paper)
    1. Semiconductor lasers. I. Title. II. Series.
TA1700.N86 2004
    621.36′6–dc22   2003060811

ISBN 0-387-40836-3   ISSN 0342-4111   Printed on acid-free paper.

© 2004 Springer-Verlag New York, Inc.
All rights reserved. This work may not be translated or copied in whole or in part without the written permission of the publisher (Springer-Verlag New York, Inc., 175 Fifth Avenue, New York, NY 10010, USA), except for brief excerpts in connection with reviews or scholarly analysis. Use in connection with any form of information storage and retrieval, electronic adaptation, computer software, or by similar or dissimilar methodology now known or hereafter developed is forbidden. The use in this publication of trade names, trademarks, service marks, and similar terms, even if they are not identified as such, is not to be taken as an expression of opinion as to whether or not they are subject to proprietary rights.

Printed in the United States of America.

9 8 7 6 5 4 3 2 1   SPIN 10944981

www.springer-ny.com

Springer-Verlag   New York Berlin Heidelberg
A member of BertelsmannSpringer Science+Business Media GmbH

Dedicated to my grandparents in the U.S.A.,

Kenichiro and Asano Kanzaki

Preface

Semiconductor lasers have been actively studied since the first laser oscillation in 1962. Through continuing efforts based on physics, characteristics of semiconductor lasers have been extensively improved. As a result, they are now widely used. For example, they are used as the light sources for bar-code readers, compact discs (CDs), CD-ROMs, magneto-optical discs (MOs), digital video discs (DVDs), DVD-ROMs, laser printers, lightwave communication systems, and pumping sources of solid-state lasers. From these facts, it may be said that semiconductor lasers are indispensable for our contemporary life.

This textbook explains the physics and fundamental characteristics of semiconductor lasers with regard to system applications. It is aimed at senior undergraduates, graduate students, engineers, and researchers. The features of this book are as follows:

1. The required knowledge to read this book is electromagnetism and introductory quantum mechanics taught in undergraduate courses. After reading this book, students will be able to understand journal papers on semiconductor lasers without difficulty.
2. To solve problems in semiconductor lasers, sometimes opposite approaches are adopted according to system applications. These approaches are compared and explained.
3. In the research of semiconductor lasers, many ideas have been proposed and tested. Some ideas persist, and others have faded out. These ideas are compared and the key points of the persisting technologies will be revealed.
4. The operating principles are often the same, although the structures seem to be different. These common concepts are essential and important; they allow us to deeply understand the physics of semiconductor lasers. Therefore, common concepts are emphasized in several examples, which will lead to both a qualitative and a quantitative understanding of semiconductor lasers.

This book consists of two parts. The first part, Chapters 1–4, reviews fundamental subjects such as the band structures of semiconductors, optical transitions, optical waveguides, and optical resonators. Based on these fundamentals, the second part, Chapters 5–8, explains semiconductor lasers.

The operating principles and basic characteristics of semiconductor lasers are discussed in Chapter 5. More advanced topics, such as dynamic single-mode lasers, quantum well lasers, and control of the spontaneous emission, are described in Chapters 6–8.

Finally, the author would like to thank Professor emeritus of the University of Tokyo, Koichi Shimoda (former professor at Keio University), Professor Kiyoji Uehara of Keio University, Professor Tomoo Fujioka of Tokai University (former professor at Keio University), and Professor Minoru Obara of Keio University for their warm encouragement and precious advice since he was a student. He is also indebted to NEC Corporation, where he started research on semiconductor lasers just after graduation from Keio University. Thanks are extended to the entire team at Springer-Verlag, especially, Mr. Frank Ganz, Mr. Frank McGuckin, Ms. Margaret Mitchell, Mr. Timothy Taylor, and Dr. Hans Koelsch, for their kind help.

Takahiro Numai
Kusatsu, Japan
September 2003

Contents

1 Band Structures

1.1 Introduction

Optical transitions, such as the *emission* and *absorption* of light, are closely related to the energies of electrons, as shown in Table 1.1. When electrons transit from high energy states to lower ones, lights are emitted, and in the reverse process, lights are absorbed. Note that *nonradiative transitions*, which do not emit lights, also exist when electrons transit from high energy states to lower ones. Light emissions, however, always accompany the transitions of electrons from high energy states to lower ones, which are referred to as *radiative transitions*.

Table 1.1. Relationship between electron energies and optical transitions

Energy of the Electrons	Optical Transition
High → low	Emission
Low → high	Absorption

Let us consider electron energies, which are the bases of the optical transitions. Figure 1.1 shows a relationship between atomic spacing and electron energies. When the atomic spacing is large, such as in gases, the electron energies are *discrete* and the *energy levels* are formed. With a decrease in the atomic spacing, the wave functions of the electrons start to overlap. Therefore, the energy levels begin to split so as to satisfy the *Pauli exclusion principle*. With an increase in the number of neighboring atoms, the number of split energy levels is enhanced, and the energy differences in the adjacent energy levels are reduced. In the semiconductor crystals, the number of atoms per cubic centimeter is on the order of 10^{22}, where the lattice constant is approximately 0.5 nm and the atomic spacing is about 0.2 nm. As a result, the spacing of energy levels is on the order of 10^{-18} eV. This energy spacing is much smaller than the *bandgap*, which is on the order of electron volts. Therefore, the constituent energy levels, which are known as the *energy bands*, are considered to be almost *continuous*.

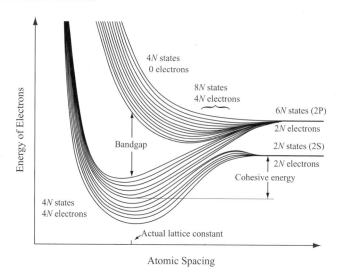

Fig. 1.1. Relationship between atomic spacing and electron energies for the diamond structure with N atoms

1.2 Bulk Structures

1.2.1 $k \cdot p$ Perturbation Theory

We study the band structures of the *bulk* semiconductors, in which constituent atoms are periodically placed in a sufficiently long range compared with the lattice spacing.

Semiconductors have carriers, such as free electrons and holes, only in the vicinity of the band edges. As a result, we would like to know the band shapes and the effective masses of the carriers near the band edges, and they often give us enough information to understand fundamental characteristics of the optical transitions. When we focus on the neighbor of the band edges, it is useful to employ the $k \cdot p$ *perturbation theory* [1–4] whose *wave vectors* ks are near the band edge wave vector k_0 inside the *Brillouin zone*. The *wave functions* and *energies* of the bands are calculated with $\Delta k = k - k_0$ as a *perturbation parameter*. For brevity, we put $k_0 = 0$ in the following.

The *Schrödinger equation* in the steady state is written as [5,6]

$$\left[-\frac{\hbar^2}{2m} \nabla^2 + V(\boldsymbol{r}) \right] \psi_{n\boldsymbol{k}}(\boldsymbol{r}) = E_n(\boldsymbol{k}) \psi_{n\boldsymbol{k}}(\boldsymbol{r}), \tag{1.1}$$

where $\hbar = h/2\pi = 1.0546 \times 10^{-34}$ J s is *Dirac's constant*, $h = 6.6261 \times 10^{-34}$ J s is *Planck's constant*, $m = 9.1094 \times 10^{-31}$ kg is the electron mass in a vacuum, $V(\boldsymbol{r})$ is a potential, $\psi_{n\boldsymbol{k}}(\boldsymbol{r})$ is a wave function, $E_n(\boldsymbol{k})$ is an *energy eigenvalue*, n is a *quantum number*, and $\boldsymbol{k}$ is a wave vector. In the *single crystals* where the atoms are placed *periodically*, the potential $V(\boldsymbol{r})$ is spatially periodic. Therefore, as a solution of (1.1), we can consider the *Bloch function* given by

$$\psi_{nk}(\boldsymbol{r}) = e^{\mathrm{i}\boldsymbol{k}\cdot\boldsymbol{r}} u_{nk}(\boldsymbol{r}), \qquad (1.2)$$

$$u_{nk}(\boldsymbol{r}) = u_{nk}(\boldsymbol{r} + \boldsymbol{R}), \qquad (1.3)$$

where $\boldsymbol{R}$ is a vector indicating the periodicity of the crystal. Equations (1.2) and (1.3) are called the *Bloch theorem*, which indicates that the wave function $u_{nk}(\boldsymbol{r})$ depends on the wave vector $\boldsymbol{k}$ and has the same periodicity as that of the crystal. Substituting (1.2) into (1.1) results in

$$\left[-\frac{\hbar^2}{2m}\nabla^2 + V(\boldsymbol{r}) + \mathcal{H}' \right] u_{nk}(\boldsymbol{r}) = E_n(\boldsymbol{k}) u_{nk}(\boldsymbol{r}), \qquad (1.4)$$

where

$$\mathcal{H}' = \frac{\hbar^2 \boldsymbol{k}^2}{2m} + \frac{\hbar}{m}\boldsymbol{k}\cdot\boldsymbol{p}, \qquad (1.5)$$

$$\boldsymbol{p} = -\mathrm{i}\,\hbar\nabla. \qquad (1.6)$$

In the $\boldsymbol{k}\cdot\boldsymbol{p}$ perturbation theory, which is only valid for small $\boldsymbol{k}$, we solve (1.4) by regarding (1.5) as the *perturbation*. Note that the name of the $\boldsymbol{k}\cdot\boldsymbol{p}$ perturbation stems from the second term on the right-hand side of (1.5).

When we consider the energy band with $n = 0$, the wave equation for the unperturbed state with $\boldsymbol{k} = 0$ is expressed as

$$\left[-\frac{\hbar^2}{2m}\nabla^2 + V(\boldsymbol{r}) \right] u_{00}(\boldsymbol{r}) = E_0(0) u_{00}(\boldsymbol{r}). \qquad (1.7)$$

In the following, for simplicity, the wave function $u_{nk}(\boldsymbol{r})$ and the energy $E_0(0)$ are represented as $u_n(\boldsymbol{k}, \boldsymbol{r})$ and E_0, respectively.

At first, we consider a nondegenerate case, in which the energy of the state n is always different from that of the other state n' ($\neq n$). From the *first-order perturbation theory* (see Appendix B), the wave function $u_0(\boldsymbol{k}, \boldsymbol{r})$ is given by

$$u_0(\boldsymbol{k}, \boldsymbol{r}) = u_0(0, \boldsymbol{r}) + \sum_{\alpha\neq 0} \frac{-\mathrm{i}\,(\hbar^2/m)\boldsymbol{k}\cdot\langle\alpha|\nabla|0\rangle}{E_0 - E_\alpha} u_\alpha(0, \boldsymbol{r}), \qquad (1.8)$$

$$\langle\alpha|\nabla|0\rangle = \int u_\alpha{}^*(0, \boldsymbol{r})\nabla u_0(0, \boldsymbol{r})\,\mathrm{d}^3\boldsymbol{r}, \qquad (1.9)$$

where $u_n(\boldsymbol{k}, \boldsymbol{r})$ is assumed to be an orthonormal function. Here, $\langle\alpha|$ and $|0\rangle$ are the *bra vector* and the *ket vector*, respectively, which were introduced by Dirac. In the *second-order perturbation theory*, an energy eigenvalue is obtained as

$$E(\boldsymbol{k}) = E_0 + \frac{\hbar^2 k^2}{2m} + \frac{\hbar^2}{m^2} \sum_{i,j} k_i k_j \sum_{\alpha \neq 0} \frac{\langle 0|p_i|\alpha\rangle\langle\alpha|p_j|0\rangle}{E_0 - E_\alpha}. \tag{1.10}$$

From (1.10), the *reciprocal effective mass tensor* is defined as

$$\left(\frac{1}{m}\right)_{ij} \equiv \frac{1}{\hbar^2}\frac{\partial^2 E}{\partial k_i \partial k_j} = \frac{1}{m}\left(\delta_{ij} + \frac{2}{m}\sum_{\alpha\neq 0}\frac{\langle 0|p_i|\alpha\rangle\langle\alpha|p_j|0\rangle}{E_0 - E_\alpha}\right). \tag{1.11}$$

With the help of (1.11), (1.10) reduces to

$$E(\boldsymbol{k}) = E_0 + \frac{\hbar^2}{2}\sum_{i,j}\left(\frac{1}{m}\right)_{ij} k_i k_j. \tag{1.12}$$

This equation includes the periodicity of the crystal (potential) in the mass of the electron as the effective mass. This effective mass is useful to make analysis easier. For example, in the quantum well (QW) structures, the electrons see both the periodic potential of the crystal and the quantum well potential. If we express equations using the effective mass, we have only to consider the quantum well potential, because the periodic potential of the crystal is already included in the effective mass. This approximation is referred to as the *effective mass approximation*.

In the following, we will consider the band structures of semiconductor crystals. Most semiconductor crystals for semiconductor lasers have a *zinc-blende structure*, in which the bottom of the conduction bands is s-orbital-like and the tops of the valence bands are p-orbital-like. In zinc-blende or diamond structures, the atomic bonds are formed via sp^3 *hybrid orbitals* as follows:

$$C : (2s)^2(2p)^2 \rightarrow (2s)^1(2p)^3$$
$$Si : (3s)^2(3p)^2 \rightarrow (3s)^1(3p)^3$$
$$ZnS : Zn : (3d)^{10}(4s)^2 \rightarrow Zn^{2-} : (3d)^{10}(4s)^1(4p)^3$$
$$S : (3s)^2(3p)^4 \rightarrow S^{2+} : (3s)^1(3p)^3$$

Therefore, the wave functions for the electrons in the zinc-blende or diamond structures are expressed as superpositions of the s-orbital function and p-orbital functions.

Let us calculate the wave functions and energies of the bands in the zinc-blende structures. We assume that both the bottom of the conduction band and the tops of the valence bands are placed at $\boldsymbol{k} = 0$, as in the direct transition semiconductors, which will be elucidated in Section 2.1. When the spin-orbit interaction is neglected, the tops of the valence bands are three-fold *degenerate* corresponding to the three p-orbitals (p_x, p_y, p_z). Here, the wave functions are written as

the s-orbital function for the bottom of the conduction band : $u_s(\boldsymbol{r})$,
the p-orbital functions for the tops of the valence bands :
$$u_x = xf(\boldsymbol{r}),\ u_y = yf(\boldsymbol{r}),\ u_z = zf(\boldsymbol{r}),\quad f(\boldsymbol{r}) : \text{a spherical function.}$$

When the energy bands are degenerate, a perturbed wave equation is given by a linear superposition of $u_s(\boldsymbol{r})$ and $u_j(\boldsymbol{r})$ $(j = x, y, z)$ as

$$u_n(\boldsymbol{k}, \boldsymbol{r}) = A u_s(\boldsymbol{r}) + B u_x(\boldsymbol{r}) + C u_y(\boldsymbol{r}) + D u_z(\boldsymbol{r}), \qquad (1.13)$$

where $A, B, C,$ and D are coefficients.

To obtain the energy eigenvalues, we rewrite (1.4) as

$$\left[-\frac{\hbar^2}{2m} \nabla^2 + V(\boldsymbol{r}) + \mathcal{H}_{d}' \right] u_n(\boldsymbol{k}, \boldsymbol{r}) = \left[E_n(\boldsymbol{k}) - \frac{\hbar^2 k^2}{2m} \right] u_n(\boldsymbol{k}, \boldsymbol{r}), \qquad (1.14)$$

$$\mathcal{H}_{d}' = \frac{\hbar}{m} \, \boldsymbol{k} \cdot \boldsymbol{p} = -\frac{i\hbar^2}{m} \boldsymbol{k} \cdot \nabla. \qquad (1.15)$$

Note that the unperturbed equation is obtained by setting $\boldsymbol{k} = 0$ in (1.14), where $E_n(0) = E_c$ and $u_0(0, \boldsymbol{r}) = u_s(\boldsymbol{r})$ for the conduction band, while $E_n(0) = E_v$ and $u_0(0, \boldsymbol{r}) = u_j(\boldsymbol{r})$ $(j = x, y, z)$ for the valence bands. Here, E_c is the energy of the bottom of the conduction band, and E_v is the energy of the tops of the valence bands.

Substituting (1.13) into (1.14); multiplying $u_s{}^*(\boldsymbol{r})$, $u_x{}^*(\boldsymbol{r})$, $u_y{}^*(\boldsymbol{r})$, and $u_z{}^*(\boldsymbol{r})$ from the left-hand side; and then integrating with respect to a volume over the space leads to

$$\begin{aligned}
(\mathcal{H}_{ss}' + E_c - \lambda) A + \mathcal{H}_{sx}' B + \mathcal{H}_{sy}' C + \mathcal{H}_{sz}' D &= 0, \\
\mathcal{H}_{xs}' A + (\mathcal{H}_{xx}' + E_v - \lambda) B + \mathcal{H}_{xy}' C + \mathcal{H}_{xz}' D &= 0, \\
\mathcal{H}_{ys}' A + \mathcal{H}_{yx}' B + (\mathcal{H}_{yy}' + E_v - \lambda) C + \mathcal{H}_{yz}' D &= 0, \\
\mathcal{H}_{zs}' A + \mathcal{H}_{zx}' B + \mathcal{H}_{zy}' C + (\mathcal{H}_{zz}' + E_v - \lambda) D &= 0,
\end{aligned} \qquad (1.16)$$

where

$$\mathcal{H}_{ij}' = \langle u_i | \mathcal{H}_{d}' | u_j \rangle = \int u_i{}^*(\boldsymbol{r}) \mathcal{H}_{d}' u_j(\boldsymbol{r}) \, \mathrm{d}^3 \boldsymbol{r} \quad (i, j = s, x, y, z), \qquad (1.17)$$

$$\lambda = E_n(\boldsymbol{k}) - \frac{\hbar^2 k^2}{2m}.$$

Note that the orthonormality of $u_s(\boldsymbol{r})$ and $u_j(\boldsymbol{r})$ $(j = x, y, z)$ were used to derive (1.16).

In (1.16), only when the determinant for the coefficients A, B, C, and D is zero, we have solutions A, B, C, and D other than $A = B = C = D = 0$. From (1.16) and (1.17), the determinant is given by

$$\begin{vmatrix}
E_c - \lambda & P k_x & P k_y & P k_z \\
P^* k_x & E_v - \lambda & 0 & 0 \\
P^* k_y & 0 & E_v - \lambda & 0 \\
P^* k_z & 0 & 0 & E_v - \lambda
\end{vmatrix} = 0, \qquad (1.18)$$

$$P = -i\frac{\hbar^2}{m} \int u_s{}^* \frac{\partial u_j}{\partial r_j} \mathrm{d}^3 \boldsymbol{r}, \quad P^* = -i\frac{\hbar^2}{m} \int u_j{}^* \frac{\partial u_s}{\partial r_j} \mathrm{d}^3 \boldsymbol{r} \qquad (1.19)$$

$$(j = x, y, z, \quad r_x = x, \ r_y = y, \ r_z = z).$$

The solutions of (1.18) are obtained as

$$E_{1,2}(\boldsymbol{k}) = \frac{E_\mathrm{c} + E_\mathrm{v}}{2} + \frac{\hbar^2 k^2}{2m} \pm \left[\left(\frac{E_\mathrm{c} - E_\mathrm{v}}{2} \right)^2 + k^2 |P|^2 \right]^{1/2}, \tag{1.20}$$

$$E_{3,4}(\boldsymbol{k}) = E_\mathrm{v} + \frac{\hbar^2 k^2}{2m}, \tag{1.21}$$

where (1.17) was used. Figure 1.2 shows the calculated results of (1.20) and (1.21). It should be noted that the spin-orbit interaction has been neglected and only the first-order perturbation has been included to derive these equations.

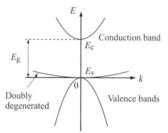

Fig. 1.2. Energy of the conduction and valence bands. Here, only the first-order perturbation is included; the spin-orbit interaction is neglected

1.2.2 Spin-Orbit Interaction

We consider the band structures by introducing the spin-orbit interaction and the second-order perturbation. First, let us treat the *spin-orbit interaction* semiclassically. As shown in Fig. 1.3, the electron with the electric charge $-e = -1.6022 \times 10^{-19}\,\mathrm{C}$ rotates about the nucleus with the electric charge $+Ze$. The velocity of the electron is $\boldsymbol{v}$, and the distance between the electron and the nucleus is $|\boldsymbol{r}|$.

Fig. 1.3. Motions of the electron

If the origin of the reference system is placed at the electron, the nucleus seems to rotate about the electron with the velocity $-\boldsymbol{v}$. As a result, due to *Biot-Savart's law*, a *magnetic flux density* $\boldsymbol{B}$ is produced at the electron, which is written as

$$\boldsymbol{B} = \frac{\mu_0}{4\pi} Ze \frac{\boldsymbol{r} \times \boldsymbol{v}}{r^3} = \frac{\mu_0}{4\pi} \frac{Ze}{m} \frac{1}{r^3} \boldsymbol{l}. \tag{1.22}$$

Here, μ_0 is magnetic permeability in a vacuum, and $\boldsymbol{l}$ is the *orbital angular momentum* given by

$$\boldsymbol{l} = \boldsymbol{r} \times \boldsymbol{p} = \boldsymbol{r} \times m\boldsymbol{v}. \tag{1.23}$$

The *spin magnetic moment* $\boldsymbol{\mu}_\mathrm{s}$ is expressed as

$$\boldsymbol{\mu}_\mathrm{s} = -\frac{e}{m}\boldsymbol{s} = -\frac{2\mu_\mathrm{B}}{\hbar}\boldsymbol{s}, \tag{1.24}$$

where $\boldsymbol{s}$ is the *spin angular momentum* and μ_B is the *Bohr magneton* defined as

$$\mu_\mathrm{B} \equiv \frac{e\hbar}{2m} = 9.2732 \times 10^{-24}\,\mathrm{A\,m^2}. \tag{1.25}$$

As a result, the *interaction energy* $\mathcal{H}_\mathrm{SO}$ between the *spin magnetic moment* $\boldsymbol{\mu}_\mathrm{s}$ and the magnetic flux density $\boldsymbol{B}$ is obtained as

$$\mathcal{H}_\mathrm{SO} = -\boldsymbol{\mu}_\mathrm{s} \cdot \boldsymbol{B} = \frac{\mu_0}{4\pi} \frac{Ze^2}{m^2} \frac{1}{r^3} \boldsymbol{l} \cdot \boldsymbol{s}. \tag{1.26}$$

From Dirac's *relativistic quantum mechanics*, the interaction energy $\mathcal{H}_\mathrm{SO}$ is given by

$$\mathcal{H}_\mathrm{SO} = \frac{\mu_0}{4\pi} \frac{Ze^2}{2m^2} \frac{1}{r^3} \boldsymbol{l} \cdot \boldsymbol{s}, \tag{1.27}$$

which is half of (1.26). As explained earlier, the spin-orbit interaction generates a magnetic field at the electron due to the orbital motions of the nucleus, and this field interacts with the electron's spin magnetic moment.

Introducing *Pauli's spin matrices* $\boldsymbol{\sigma}$ such as

$$\boldsymbol{s} = \frac{\hbar}{2} \boldsymbol{\sigma}, \tag{1.28}$$

$$\sigma_x = \begin{bmatrix} 0 & 1 \\ 1 & 0 \end{bmatrix}, \quad \sigma_y = \begin{bmatrix} 0 & -i \\ i & 0 \end{bmatrix}, \quad \sigma_z = \begin{bmatrix} 1 & 0 \\ 0 & -1 \end{bmatrix}, \tag{1.29}$$

we can write the *spin-orbit interaction Hamiltonian* $\mathcal{H}_\mathrm{SO}$ as

$$\mathcal{H}_\mathrm{SO} = \frac{\mu_0}{4\pi} \frac{Ze^2}{2m^2} \frac{1}{r^3} \frac{\hbar}{2} \boldsymbol{l} \cdot \boldsymbol{\sigma}. \tag{1.30}$$

If we express the *up-spin* ↑ ($s_z = \hbar/2$) as α and the *down-spin* ↓ ($s_z = -\hbar/2$) as β, they are written in matrix form as

$$\alpha = \begin{bmatrix} 1 \\ 0 \end{bmatrix}, \quad \beta = \begin{bmatrix} 0 \\ 1 \end{bmatrix}. \tag{1.31}$$

Using α and β, we obtain the following relations:

$$\sigma_z \alpha = \alpha, \quad \sigma_z \beta = -\beta. \tag{1.32}$$

To treat the spin-orbit interaction, it is convenient to use the *spherical polar coordinate systems*. Therefore, we rewrite the spin-orbit interaction Hamiltonian $\mathcal{H}_{\mathrm{SO}}$ as

$$\mathcal{H}_{\mathrm{SO}} = \frac{\hbar}{2}\, \xi(\boldsymbol{r})\, \boldsymbol{l} \cdot \boldsymbol{\sigma} = \frac{\hbar}{2}\, \xi(\boldsymbol{r}) \left(l_z \sigma_z + \frac{l_+ \sigma_- + l_- \sigma_+}{2} \right), \tag{1.33}$$

where

$$\xi(\boldsymbol{r}) = \frac{\mu_0}{4\pi} \frac{Ze^2}{2m^2} \frac{1}{r^3},$$
$$l_+ = l_x + \mathrm{i}\, l_y, \quad l_- = l_x - \mathrm{i}\, l_y, \tag{1.34}$$
$$\sigma_+ = \sigma_x + \mathrm{i}\, \sigma_y, \quad \sigma_- = \sigma_x - \mathrm{i}\, \sigma_y.$$

When this $\mathcal{H}_{\mathrm{SO}}$ is added to the perturbation term, the wave equation (1.14) is modified as

$$\left[-\frac{\hbar^2}{2m} \nabla^2 + V(\boldsymbol{r}) + \mathcal{H}'_{\mathrm{d}} + \mathcal{H}_{\mathrm{SO}} \right] u_n(\boldsymbol{k}, \boldsymbol{r}) = \left[E_n(\boldsymbol{k}) - \frac{\hbar^2 k^2}{2m} \right] u_n(\boldsymbol{k}, \boldsymbol{r}). \tag{1.35}$$

It should be noted that $\boldsymbol{l}$ operates on $\mathrm{e}^{\mathrm{i}\boldsymbol{k}\cdot\boldsymbol{r}}$ in the Bloch function, but this operation is neglected because the result is much smaller than the other terms.

To solve (1.35), it is useful to represent the wave functions $u_n(\boldsymbol{k}, \boldsymbol{r})$ in the spherical polar coordinate systems such as

$$u_s = u_s,$$

$$u_+ = -\frac{u_x + u_y}{\sqrt{2}} \sim -\frac{x + y}{\sqrt{2}},$$

$$u_- = \frac{u_x - u_y}{\sqrt{2}} \sim \frac{x - y}{\sqrt{2}}, \tag{1.36}$$

$$u_z \sim z.$$

In (1.36), the spherical function $f(\boldsymbol{r})$ is omitted after $\sim$ to simplify expressions. Note that $\sqrt{2}$ in the denominators is introduced to normalize the wave functions. Using the spherical harmonic function Y_l^m, the wave functions u_+, u_-, and u_z are also expressed as

$$u_+ = Y_1^1 = -\frac{1}{2}\sqrt{\frac{3}{2\pi}}\frac{x+iy}{\sqrt{x^2+y^2+z^2}} = -\frac{1}{2}\sqrt{\frac{3}{2\pi}}e^{i\phi}\sin\theta,$$

$$u_- = Y_1^{-1} = \frac{1}{2}\sqrt{\frac{3}{2\pi}}\frac{x-iy}{\sqrt{x^2+y^2+z^2}} = \frac{1}{2}\sqrt{\frac{3}{2\pi}}e^{-i\phi}\sin\theta, \qquad (1.37)$$

$$u_z = Y_1^0 = \frac{1}{2}\sqrt{\frac{3}{\pi}}\frac{z}{\sqrt{x^2+y^2+z^2}} = \frac{1}{2}\sqrt{\frac{3}{\pi}}\cos\theta,$$

where $x = r\sin\theta\cos\phi$, $y = r\sin\theta\sin\phi$, and $z = r\cos\theta$.

When we consider the up- and down-spins α and β, we have eight wave functions as follows:

$$u_s\alpha, \quad u_s\beta, \quad u_+\alpha, \quad u_+\beta, \quad u_z\alpha, \quad u_z\beta, \quad u_-\alpha, \quad u_-\beta.$$

Therefore, we have to calculate the elements of the 8×8 matrix to obtain the energy eigenvalues from (1.35).

For brevity, we assume that $\boldsymbol{k}$ is directed in the z-direction and put

$$k_z = k, \quad k_x = k_y = 0. \qquad (1.38)$$

In this case, however, we have only to solve the determinant for the 4×4 matrix on $(u_s\alpha, \ u_+\beta, \ u_z\alpha, \ u_-\beta)$ or $(u_s\beta, \ u_-\alpha, \ u_z\beta, \ u_+\alpha)$ because of the symmetry in the 8×8 matrix. This determinant for the 4×4 matrix is written as

$$\begin{vmatrix} E_{\rm c} - \lambda & 0 & Pk & 0 \\ 0 & E_{\rm v} - \lambda - \dfrac{\Delta_0}{3} & \dfrac{\sqrt{2}}{3}\Delta_0 & 0 \\ P^*k & \dfrac{\sqrt{2}}{3}\Delta_0 & E_{\rm v} - \lambda & 0 \\ 0 & 0 & 0 & E_{\rm v} - \lambda + \dfrac{\Delta_0}{3} \end{vmatrix} = 0, \qquad (1.39)$$

where the terms including Δ_0 are the matrix elements of $\mathcal{H}_{\rm SO}$, and the other terms are those of $\mathcal{H}_{\rm d}'$. Here, using $\xi(\boldsymbol{r})$ in (1.34), Δ_0 is defined as

$$\frac{\Delta_0}{3} = \frac{\hbar^2}{2}\int u_+{}^* u_+ \xi(\boldsymbol{r})\,{\rm d}^3\boldsymbol{r} = \frac{\hbar^2}{2}\int u_-{}^* u_- \xi(\boldsymbol{r})\,{\rm d}^3\boldsymbol{r}$$

$$= \frac{\hbar^2}{4}\int (u_x{}^2 + u_y{}^2)\xi(\boldsymbol{r})\,{\rm d}^3\boldsymbol{r} = \frac{\hbar^2}{2}\int u_z{}^2\,\xi(\boldsymbol{r})\,{\rm d}^3\boldsymbol{r}. \qquad (1.40)$$

From (1.39), the energy of valence band 1 is obtained as

$$E_{\rm v1}(\boldsymbol{k}) = E_{\rm v} + \frac{\Delta_0}{3} + \frac{\hbar^2 k^2}{2m}. \qquad (1.41)$$

When $|P|^2 k^2$ is small enough, the energy of the conduction band E_c reduces to

$$E_c(\boldsymbol{k}) = E_c + \frac{\hbar^2 k^2}{2m} + \frac{|P|^2 k^2}{3} \left(\frac{2}{E_g} + \frac{1}{E_g + \Delta_0} \right), \qquad (1.42)$$

where

$$E_g = E_c - E_v - \frac{\Delta_0}{3}. \qquad (1.43)$$

Similarly, the energies of valence bands 2 and 3 are given by

$$E_{v2}(\boldsymbol{k}) = E_v + \frac{\Delta_0}{3} + \frac{\hbar^2 k^2}{2m} - \frac{2|P|^2 k^2}{3E_g}, \qquad (1.44)$$

$$E_{v3}(\boldsymbol{k}) = E_v - \frac{2}{3}\Delta_0 + \frac{\hbar^2 k^2}{2m} - \frac{|P|^2 k^2}{3(E_g + \Delta_0)}. \qquad (1.45)$$

These results, which were obtained under the first-order $\boldsymbol{k} \cdot \boldsymbol{p}$ perturbation, are shown in Fig. 1.4. From the definition of effective mass in (1.11), the band with energy $E_{v1}(\boldsymbol{k})$ is referred to as the *heavy hole band*, and that with $E_{v2}(\boldsymbol{k})$ is called the *light hole band*. It should be noted that the heavy hole band and the light hole band are degenerate at a point $\boldsymbol{k} = 0$. The band with energy $E_{v3}(\boldsymbol{k})$ is designated as the *split-off band*, and Δ_0 is called the *split-off energy*.

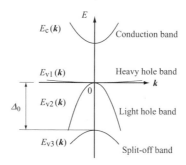

Fig. 1.4. Energy bands when the spin-orbit interaction is considered under the first-order perturbation

If we consider the second-order perturbation, the energies of the valence bands are given by

$$E_{v1,2}(\boldsymbol{k}) = E_v + \frac{\Delta_0}{3} + A_2 k^2$$
$$\pm \left[B_2^2 k^4 + C_2^2 \left(k_x^2 k_y^2 + k_y^2 k_z^2 + k_z^2 k_x^2 \right) \right]^{1/2} \qquad (1.46)$$
$$(1 \rightarrow +, 2 \rightarrow -)$$

$$E_{v3}(\boldsymbol{k}) = E_v - \frac{2}{3}\Delta_0 + A_2 k^2. \qquad (1.47)$$

The coefficients A_2, B_2, and C_2 in (1.46) and (1.47) are experimentally determined by the *cyclotron resonance* (see Appendix A). When the second-order perturbation is included, all the valence bands become upward-convex, as shown in Fig. 1.5, but degeneracy of the heavy hole band and the light hole band at $\boldsymbol{k} = 0$ remains.

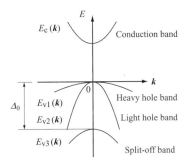

Fig. 1.5. Energy bands when the spin-orbit interaction is considered under the second-order perturbation

The preceding analysis treats the *direct transition* semiconductors where both the bottom of the conduction band and the tops of the valence bands are placed at $\boldsymbol{k} = 0$. In the *indirect transition* semiconductors, $\boldsymbol{k}$ of the bottom of the conduction band and that of the tops of the valence bands are different. It should also be noted that the effective masses depend on the direction of $\boldsymbol{k}$. Therefore, the band structures are more complicated.

Let us consider the wave functions of the valence bands under the second-order perturbation. Due to the spin-orbit interaction, the quantum states are indicated by $\boldsymbol{j} = \boldsymbol{l} + \boldsymbol{s}$ where $\boldsymbol{l}$ is the angular momentum operator and $\boldsymbol{s}$ is the spin operator. Therefore, as indexes of the wave functions, we use the quantum numbers j and m_j, which represent the eigenvalues of the operators $\boldsymbol{j}$ and j_z, respectively. The relation between the operators and the eigenvalues is summarized in Table 1.2.

Table 1.2. Relation between the operators and eigenvalues

Operator	Eigenvalue
l^2	$l(l+1)\hbar^2$ ($l = 0$: s-orbital, $l = 1$: p-orbitals)
l_z	$m_l \hbar$, $m_l = 1, 0, -1$
s^2	$s(s+1)\hbar^2$, $s = 1/2$
s_z	$m_s \hbar$, $m_s = 1/2, -1/2$
j^2	$j(j+1)\hbar^2$, $j = 3/2, 1/2$
j_z	$m_j \hbar$, $m_{j=3/2} = 3/2, 1/2, -1/2, -3/2$, $m_{j=1/2} = 1/2, -1/2$

When we express the wave functions as $|j, m_j\rangle$, the wave functions are expressed as:

for the *heavy hole band*

$$\left|\frac{3}{2}, \frac{3}{2}\right\rangle = \frac{1}{\sqrt{2}}\,|(x + \mathrm{i}\,y)\alpha\rangle,$$

$$\left|\frac{3}{2}, -\frac{3}{2}\right\rangle = \frac{1}{\sqrt{2}}\,|(x - \mathrm{i}\,y)\beta\rangle,$$

(1.48)

for the *light hole band*

$$\left|\frac{3}{2}, \frac{1}{2}\right\rangle = \frac{1}{\sqrt{6}}\,|2z\alpha + (x + \mathrm{i}\,y)\beta\rangle,$$

$$\left|\frac{3}{2}, -\frac{1}{2}\right\rangle = \frac{1}{\sqrt{6}}\,|2z\beta - (x - \mathrm{i}\,y)\alpha\rangle,$$

(1.49)

and for the *split-off band*

$$\left|\frac{1}{2}, \frac{1}{2}\right\rangle = \frac{1}{\sqrt{3}}\,|z\alpha - (x + \mathrm{i}\,y)\beta\rangle,$$

$$\left|\frac{1}{2}, -\frac{1}{2}\right\rangle = \frac{1}{\sqrt{3}}\,|z\beta + (x - \mathrm{i}\,y)\alpha\rangle.$$

(1.50)

1.3 Quantum Structures

1.3.1 Potential Well

The semiconductor structures whose sizes are small enough that their *quantum effects* may be significant are called *quantum structures*.

The electrons in the quantum structures see both the *periodic potential*, corresponding to the periodicity of the crystals, and the quantum well potential. Before studying the energy bands in the quantum structures, we will review the energies and wave functions of a particle in a square well potential.

Here, we assume that a carrier exists in a square potential well, as shown in Fig. 1.6.

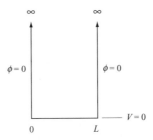

Fig. 1.6. Square well potential

The square potential $V(r)$ is

$$\left.\begin{array}{l} V(r) = 0 \ \text{ inside the well} \\ V(r) = \infty \ \text{outside the well} \end{array}\right\} . \tag{1.51}$$

Note that the potential $V(r)$ is not periodic. When the potential well is a cube with a side L, the boundary conditions for a wave function $\phi(x, y, z)$ are given by

$$\left.\begin{array}{l} \phi(0, y, z) = \phi(L, y, z) = 0 \\ \phi(x, 0, z) = \phi(x, L, z) = 0 \\ \phi(x, y, 0) = \phi(x, y, L) = 0 \end{array}\right\} . \tag{1.52}$$

Under these boundary conditions, the wave function $\phi(x, y, z)$ and the energy E are obtained as

$$\phi(x, y, z) = \sqrt{\frac{8}{L^3}} \sin k_x x \cdot \sin k_y y \cdot \sin k_z z, \quad E = \frac{\hbar^2}{2m}({k_x}^2 + {k_y}^2 + {k_z}^2),$$

$$k_x = \frac{n_x \pi}{L}, \ k_y = \frac{n_y \pi}{L}, \ k_z = \frac{n_z \pi}{L} \quad (n_x, n_y, n_z = 1, 2, 3, \cdots).$$

$$\tag{1.53}$$

Figure 1.7 shows the wave functions ϕs and the energies Es for a one-dimensional square well potential. As found in (1.53), the energies E are *discrete* and their values are proportional to a square of the quantum number n_x. Also, with a decrease in L, an energy separation between the energy levels increases. The wave functions can take negative values as well as positive ones. The squares of the wave functions show possibilities of existence, so negative values are also allowed for the wave functions.

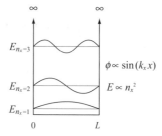

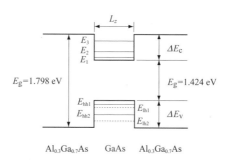

Fig. 1.7. Wave functions ϕs and energies Es in a one-dimensional square well potential

1.3.2 Quantum Well, Wire, and Box

First, we define some technical terms. Figure 1.8 shows the energies of the conduction band and valence bands for GaAs sandwiched by AlGaAs at a point $\boldsymbol{k} = 0$.

Fig. 1.8. Quantum well structure

The low energy regions for the electrons in the conduction band and the holes in the valence band are called *potential wells*. Note that, in Fig. 1.8, the vertical line shows the energy of the electrons, and the energy of the holes decreases with an increase in the height of the vertical line. In this figure, the potential well for the electrons in the conduction band and that for the holes in the valence band are both GaAs. When the width of this potential well L_z is on the order of less than several tens of nanometers, this well is referred to as the *quantum well*. The bandgaps of AlGaAs layers placed at both sides of GaAs are higher than that of GaAs. As a result, these AlGaAs layers function as the energy barriers for GaAs, and they are designated as the *energy barrier layers*. At the interfaces of the quantum well and the barriers, there are the energy difference in the conduction bands ΔE_c and that in the valence bands ΔE_v, which are called the *band offsets*.

The periods of the potential for the semiconductor crystals are the lattice constants, which are on the order of 0.5 nm. In contrast, the thickness of the potential wells or the barriers in the quantum structures is between the order of nanometers and several tens of nanometers. Hence, in the quantum structures, the electrons and the holes see both the *periodic potential* and the *quantum potential*. If we use the *effective mass*, the effect of the periodic potential is included in the effective mass, as shown in (1.12), and we have only to consider the quantum potential, which is referred to as the *effective mass approximation*.

Under the effective mass approximation, a wave function in the quantum structure is obtained by a product of the *base function* ψ and the *envelope function* ϕ. As the base function, we use a wave function for the periodic potential

$$\psi_{n\boldsymbol{k}}(\boldsymbol{r}) = e^{i\boldsymbol{k}\cdot\boldsymbol{r}} u_{n\boldsymbol{k}}(\boldsymbol{r}), \quad u_{n\boldsymbol{k}}(\boldsymbol{r}) = u_{n\boldsymbol{k}}(\boldsymbol{r}+\boldsymbol{R}). \tag{1.54}$$

As the envelope function, we use a wave function for the quantum potential. For example, for a cube with the potential shown in Fig. 1.6, ϕ is given by

$$\phi(x,y,z) = \sqrt{\frac{8}{L^3}} \sin k_x x \cdot \sin k_y y \cdot \sin k_z z. \tag{1.55}$$

(a) One-Dimensional Quantum Well

Let us consider a sheet with side lengths of L_x, L_y, and L_z. As shown in Fig. 1.9, we assume that only L_z is a quantum size, which satisfies $L_z \ll L_x, L_y \approx L$. Such a structure is called a *one-dimensional quantum well*.

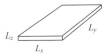

Fig. 1.9. One-dimensional quantum well

The energies of the carriers are written as

$$E = E_{xy} + E_z,$$
$$E_{xy} = \frac{\hbar^2}{2m^*}\frac{\pi^2}{L^2}(n_x{}^2 + n_y{}^2), \quad E_z = \frac{\hbar^2}{2m^*}\frac{\pi^2}{L_z{}^2}n_z{}^2. \tag{1.56}$$

Here, $\hbar$ is Dirac's constant; m^* is the effective mass of the carrier; and n_x, n_y, and n_z are quantum numbers. If n_x, n_y, and n_z are of the same order, we have $E_{xy} \ll E_z$.

Figure 1.10 schematically shows the energies of the valence bands in the one-dimensional quantum well. In this figure, E_{hh1} and E_{hh2} (solid lines) represent the heavy hole bands, and E_{hh1} and E_{lh2} (broken lines) express the light hole bands. Here, subscripts 1 and 2 are the quantum numbers n_zs. As shown in Fig. 1.10, the quantum well structures remove degeneracy of the heavy hole band and the light hole band at a point $\boldsymbol{k} = 0$, because the potential symmetry of the quantum wells is lower than that of the bulk structures.

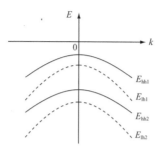

Fig. 1.10. Valence bands in a one-dimensional quantum well

Let us calculate the *density of states* in the one-dimensional quantum well. As an example, we treat the density of states for $n_z = 1$. The density of states is determined by combinations of n_x and n_y. When n_x and n_y are large enough, the combinations (n_x, n_y) for a constant energy E_{xy} are represented by the points on the circumference of a circle with a radius r, which is given by

$$r^2 = n_x^{\,2} + n_y^{\,2} = \frac{2m^* L^2}{\hbar^2 \pi^2} E_{xy}. \tag{1.57}$$

Because both n_x and n_y are positive numbers, the number S of the combinations (n_x, n_y) is given by the area of the quarter circle with the radius r, as shown in Fig. 1.11.

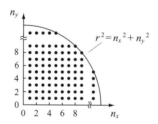

Fig. 1.11. Combinations of n_x and n_y

From Fig. 1.11, S is expressed as

$$S = \frac{1}{4}\pi r^2 = \frac{\pi}{4}(n_x{}^2 + n_y{}^2) = \frac{\pi}{4}\frac{2m^* L^2}{\hbar^2 \pi^2} E_{xy} = \frac{m^* L^2}{2\hbar^2 \pi} E_{xy}. \quad (1.58)$$

Considering the up- and down-spins, the number of states N is written as

$$N = 2S = \frac{m^* L^2}{\hbar^2 \pi} E_{xy}. \quad (1.59)$$

Substituting $E_{xy} = E - E_{z=1}$ into (1.59), the electron concentration n for the energy between zero and E is obtained as

$$n = \frac{N}{L^2 L_z} = \frac{m^*}{\hbar^2 \pi L_z}(E - E_{z=1}). \quad (1.60)$$

When we define the density of states for the energy between E and $E + \mathrm{d}E$ as $\rho_1(E)$, we have

$$\int \rho_1(E)\,\mathrm{d}E \equiv n. \quad (1.61)$$

From (1.60) and (1.61), we obtain

$$\rho_1(E) \equiv \frac{\mathrm{d}n}{\mathrm{d}E} = \frac{m^*}{\hbar^2 \pi L_z}. \quad (1.62)$$

Similarly, the densities of states for $n_z = 2, 3, \cdots$ are calculated, and the results are shown in Fig. 1.12.

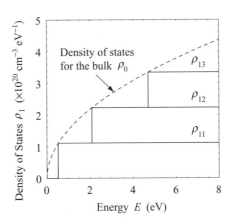

Fig. 1.12. Density of states for the one-dimensional quantum well for $L_z = 3$ nm and $m^* = 0.08m$ (solid line) and that for the bulk structures (broken line)

In Fig. 1.12, L_z is 3 nm, m^* is $0.08m$ where m is the electron mass in a vacuum, and $\rho_1(E)$ for $n_z = 1$, 2, 3 are indicated as ρ_{11}, ρ_{12}, ρ_{13}, respectively. It should be noted that the density of states for the one-dimensional quantum well is a *step function*. In contrast, the bulk structures have the density of states such as

$$\rho_0(E) = \frac{(2m^*)^{3/2}}{2\pi^2\hbar^3}E^{1/2}, \tag{1.63}$$

which is proportional to $E^{1/2}$ as shown by a broken line, because the number of states is the volume of $1/8$ of the sphere with the radius r.

(b) Two-Dimensional Quantum Well (Quantum Wire)

A stripe with $L_x \gg L_y, L_z$, shown in Fig. 1.13, is designated the *two-dimensional quantum well* or the *quantum wire*. Note that L_y and L_z are quantum sizes.

Fig. 1.13. Two-dimensional quantum well (quantum wire)

For brevity, if we put $L_y = L_z = L$, the energies are written as

$$E = E_x + E_{yz},$$
$$E_x = \frac{\hbar^2}{2m^*}\frac{\pi^2}{L_x^2}n_x{}^2, \quad E_{yz} = \frac{\hbar^2}{2m^*}\frac{\pi^2}{L^2}(n_y{}^2 + n_z{}^2). \tag{1.64}$$

For a pair of quantum numbers (n_y, n_z), the density of states $\rho_2(E)$ is obtained as

$$\rho_2(E) = \frac{\sqrt{2m^*}}{\hbar\pi L^2}E_x{}^{-1/2} = \frac{\sqrt{2m^*}}{\hbar\pi L^2}(E - E_{yz})^{-1/2}. \tag{1.65}$$

Figure 1.14 shows the calculated result of (1.65) with $L_z = 3$ nm and $m^* = 0.08m$, which are the same as in Fig. 1.12. When the energy E is equal to E_{yz}, the density of states $\rho_2(E)$ is infinity. When E exceeds E_{yz}, $\rho_2(E)$ decreases in proportion to $(E - E_{yz})^{-1/2}$. As a result, the density of states $\rho_2(E)$ has a saw-toothed shape.

(c) Three-Dimensional Quantum Well (Quantum Box)

As shown in Fig. 1.15, a box whose L_x, L_y, and L_z are all quantum sizes, is named the *three-dimensional quantum well* or the *quantum box*.

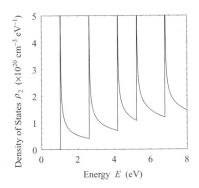

Fig. 1.14. Density of states for the two-dimensional quantum well (quantum wire)

Fig. 1.15. Three-dimensional quantum well (quantum box)

For brevity, if we put $L_x = L_y = L_z = L$, the energies are written as

$$E = E_x + E_y + E_z,$$

$$E_x = \frac{\hbar^2}{2m^*} \frac{\pi^2}{L^2} n_x{}^2, \quad E_y = \frac{\hbar^2}{2m^*} \frac{\pi^2}{L^2} n_y{}^2, \quad E_z = \frac{\hbar^2}{2m^*} \frac{\pi^2}{L^2} n_z{}^2. \tag{1.66}$$

It should be noted that the energies are completely discrete. The density of states $\rho_3(E)$ is a *delta function*, which is written as

$$\rho_3(E) = 2 \sum_{n_x, n_y, n_z} \delta(E - E_x - E_y - E_z). \tag{1.67}$$

Figure 1.16 shows the number of states per volume and the density of states in the three-dimensional quantum well.

The energy distributions of the electrons are given by the product of the densities of states and the Fermi-Dirac distribution functions. With an increase in the dimension of the quantum wells, the energy bandwidths of the densities of states decrease. Therefore, the energy distribution of the electron concentrations narrows with an increase in the dimension of the quantum wells, as shown in Fig. 1.17.

As explained earlier, the energy distribution of the electrons in the quantum structures is narrower than that in the bulk structures. Therefore, the optical gain concentrates on a certain energy (wavelength). As a result, in the quantum well lasers, a low threshold current, a high speed modulation, a low chirping, and a narrow spectral linewidth are expected, which will be described in Chapter 7.

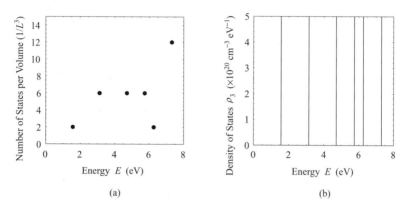

Fig. 1.16. (a) Number of states per unit volume and (b) the density of states for the three-dimensional quantum well (quantum box)

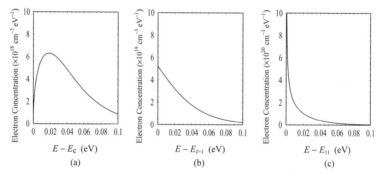

Fig. 1.17. Energy distribution of electron concentrations in quantum wells: (a) bulk structure, (b) 1-D quantum structure, and (c) 2-D quantum structure

1.4 Super Lattices

In the previous section, we studied quantum structures. Here, we consider *super lattices*, which include array quantum structures and solitary ones. From the viewpoints of the potential and the period, super lattices are classified as follows.

1.4.1 Potential

Figure 1.18 shows three kinds of super lattices. In this figure, the horizontal direction indicates the position of the layers, and the vertical direction represents the energy of the electrons. As a result, with an increase in the height of the vertical direction, the energy of electrons increases and that of the holes decreases. As shown in Fig. 1.18 (a), in *Type I* super lattice, a spatial position of the potential well for the electrons in the conduction band is the same as that for the holes in the valence band. Therefore, both electrons and the holes are confined to semiconductor layer B, which has a narrower bandgap than layer A. In *Type II* super lattice in Fig. 1.18 (b), the electrons in the conduction band are confined to semiconductor layer B, and the holes in the valence band are confined to semiconductor layer A. In *Type III* super lattice in Fig. 1.18 (c), the energy of the conduction band of semiconductor layer B overlaps that of the valence band of layer A, which results in the *semimetal*. Note that in some articles, Type II and Type III are called Type I' and Type II, respectively. The names other than Type I may be different, but the important point is that the characteristics of the super lattices are highly dependent on the shapes of the potentials.

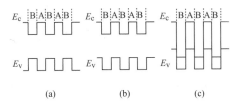

(a)   (b)   (c)

Fig. 1.18. Classification of super lattices by potential: (a) Type I, (b) Type II, and (c) Type III

1.4.2 Period

The characteristics of the super lattices also depend on the periods of the constituent layers. Figure 1.19 shows the relationships between the characteristics of the super lattices and the thickness of the barriers and wells. When each layer thickness is larger than several tens of nanometers, only the bulk characteristics are observed. If the barrier thickness is less than several tens of nanometers, the *quantum mechanical tunneling effect* appears, which leads to tunnel diodes (Esaki diodes) or devices using the *resonant tunneling effect*. Although the barriers are thick and only the wells are thin, quantum energy levels are formed in the wells. If such wells are used as the active layers in the light emitting devices, narrow light emission spectra are obtained. When both the barriers and the wells are thinner than the order of 10 nm, the wave functions of a well start to penetrate adjacent wells. As a result, the wave functions of each well overlap with each other, which produces the *minizones* and induces the *Bloch oscillations* or the *negative resistances*. As the thickness of both the barriers and the wells decreases further down to the order of atomic layers, *bending of the Brillouin zones* appears, which will transform the indirect transition materials into the direct transition ones.

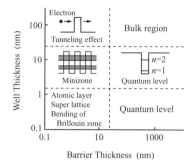

Fig. 1.19. Classification of super lattices by period

1.4.3 Other Features in Addition to Quantum Effects

In order to fabricate the quantum structures, barriers and wells are required. Because the barriers and wells must have different bandgaps, different kinds of semiconductor materials are needed. Therefore, the quantum structures are inevitably *heterostructures*.

To achieve a low threshold current and a high light emission efficiency in *semiconductor lasers*, both the carriers and the light should be confined to the active layers where the light is generated and amplified. Therefore, the *double heterostructure*, in which the heterostructures are placed at both sides of the active layer, is adopted in semiconductor lasers. Figure 1.20 shows the electron energies and refractive indexes of the double heterostructure. Because the energy barriers exist in the junction boundaries, the carriers are confined to well layer B. In addition, the semiconductors with larger bandgaps generally have smaller refractive indexes. Therefore, light is confined to well layer B. As a result, both the carriers and the light are confined to well layer B, which is used as the active layer.

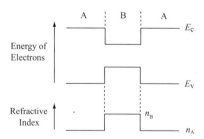

Fig. 1.20. Energies and refractive indexes of the double heterostructure

Finally, we consider a layer epitaxially grown on the semiconductor substrate whose lattice constant is different from that of the grown layer. When the layer thickness exceeds the *critical thickness*, the grown layer is plasti-cally deformed and the *dislocations* are induced in it. If the dislocations are generated, the carriers are dissipated without emitting light. Consequently, characteristics of light emitting devices become extremely low.

Although the layer thickness is thinner than the critical thickness, the grown layers are elastically deformed and the dislocations are not generated in the grown layers. Due to the *elastic strains*, the atomic spacings of the grown layers change, which modifies the band structure of the grown layer. This technology is referred to as *band-structure engineering* and attracts a lot of attention. Because the quantum structures have thin layers, they are suitable for band-structure engineering using elastic strains, and they improve characteristics of semiconductor lasers, which will be explained in Chapter 7.

2 Optical Transitions

2.1 Introduction

Among energy states, the state with the lowest energy is most stable. Therefore, the electrons in semiconductors tend to stay in low energy states. If they are excited by thermal energy, light, or electron beams, the electrons absorb these energies and transit to high energy states. These *transitions* of the electrons from low energy states to high energy states are called *excitations*. High energy states, however, are unstable. As a result, to take stable states, the electrons in high energy states *transit* to low energy states in certain *lifetimes*. These transitions of the excited electrons from high energy states to low energy states are referred to as *relaxations*. The excitation and relaxation processes between the valence band and the conduction band are shown in Fig. 2.1.

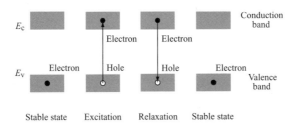

Fig. 2.1. Excitation and relaxation

In semiconductors, the transitions of electrons from high energy states to low energy states are designated *recombinations* of the electrons and the holes. In the recombinations of the electrons and the holes, there are *radiative recombinations* and *nonradiative recombinations*. The radiative recombinations emit *photons*, and the energies of the photons correspond to a difference in the energies between the initial and final energy states related to the transitions. In contrast, in the nonradiative recombinations, the *phonons* are emitted to crystal lattices or the electrons are trapped in the defects, and the transition energy is transformed into forms other than light. The *Auger processes* are also categorized as nonradiative recombinations. To obtain high efficiency semiconductor light emitting devices, we have to minimize the non-radiative recombinations. However, to enhance modulation characteristics, the nonradiative recombination centers may be intentionally induced in the active layers, because they reduce the carrier lifetimes (see Section 5.1).

Let us consider the transitions of the electrons from the bottom of the conduction band to the top of the valence band. A semiconductor, in which the bottom of the conduction band and the top of the valence band are placed at a common wave vector k, is the *direct transition* semiconductor. A semiconductor, in which the bottom of the conduction band and the top of the valence band have different k-values, is the *indirect transition* semiconductor. These direct and indirect transitions are schematically shown in Fig. 2.2. In transitions of the electrons, the energy and the momentum are conserved, respectively. Therefore, the phonons do not take part in direct transitions. Because the wave vector k of the phonons is much larger than that of the photons, the phonon transitions accompany the indirect transitions to satisfy the momentum conservation law. Hence, in the direct transitions, the transition probabilities are determined by only the electron transition probabilities. In contrast, in the indirect transitions, the transition probabilities are given by a product of the electron transition probabilities and the phonon transition probabilities. As a result, the transition probabilities of the direct transitions are much higher than those of the indirect transitions. Consequently, the direct transition semiconductors are superior to the indirect ones for light emitting devices.

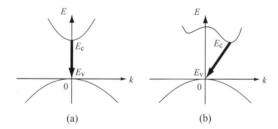

Fig. 2.2. (a) Direct and (b) indirect transition semiconductors

2.2 Light Emitting Processes

Light emission due to the radiative recombinations is called the *luminescence*. According to the lifetime, the excitation methods, and the energy states related to the transitions, light emitting processes are classified as follows.

2.2.1 Lifetime

With regard to the lifetime, there are two light emissions: *fluorescence*, with a short lifetime of 10^{-9}–10^{-3} s, and *phosphorescence*, with a long lifetime of 10^{-3} s to one day.

2.2.2 Excitation

Luminescence due to optical excitation (pumping) is *photoluminescence*, which is widely used to characterize materials. Optical excitation is also used to pump dye lasers (for example, Rhodamine 6G and Coumalin) and solid-state lasers (for example, YAG and ruby). When the photon energy of the pumping light is $\hbar\omega_1$ and that of the luminescence is $\hbar\omega_2$, the luminescence with $\hbar\omega_2 < \hbar\omega_1$ is called *Stokes luminescence* and that with $\hbar\omega_2 > \hbar\omega_1$ is designated *anti-Stokes luminescence*. Luminescence caused by electrical excitation is *electroluminescence*, which has been used for panel displays. In particular, luminescence by current injection is called *injection-type electroluminescence*; it has been used for light emitting diodes (LEDs) and semiconductor lasers or laser diodes (LDs). In such injection-type optical devices, the carriers are injected into the active layers by forward bias across the pn junctions. Note that the current (carrier) injection is also considered the excitation, because it generates a lot of high energy electrons. The luminescence due to electron beam irradiation is *cathodoluminescence*, which has been adopted to characterize materials. The luminescence induced by mechanical excitation using stress is *triboluminescence*, and that by thermal excitation is *thermoluminescence*. Luminescence during a chemical reaction is referred to chemiluminescence; it has not been reported in semiconductors.

2.2.3 Transition States

Figure 2.3 shows light emission processes between various energy states. They are classified into impurity recombinations, interband recombinations, and exciton recombinations.

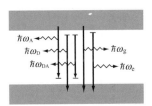

Fig. 2.3. Light emission processes

In *impurity recombinations*, there is recombination between the electron in the conduction band and the empty acceptor level with the photon energy of $\hbar\omega_A$, recombination between the electron in the donor level and the hole in the valence band with the photon energy of $\hbar\omega_D$, and recombination between the electron in the donor level and the empty acceptor level with the photon energy of $\hbar\omega_{DA}$. By observing light emissions due to these recombinations at extremely low temperatures, information on doping characteristics is obtained.

In *interband recombinations* between the conduction and valence bands, light emission in the vicinity of the band edges with the photon energy of $\hbar\omega_g$ is dominant. This *band edge emission* is used in most LEDs and LDs composed of III-V group semiconductors.

The *exciton recombination*, with the photon energy of $\hbar\omega_e$, is the recombination of the electron generated by decay of an exciton and the hole in the valence band.

2.3 Spontaneous Emission, Stimulated Emission, and Absorption

Figure 2.4 schematically shows *radiations* and *absorption*. In the radiations, there are *spontaneous emission* and *stimulated emission* (or *induced emission*). As shown in Fig. 2.4 (a), spontaneous emission is a radiative process in which an excited electron decays in a certain lifetime and a photon is emitted. Note that spontaneous emission takes place irrespective of incident lights. In contrast, in the stimulated emission illustrated in Fig. 2.4 (b), an incident light induces a radiative transition of an excited electron. The emitted light due to the stimulated emission has the same *wavelength*, *phase*, and *direction* as the incident light. Therefore, the light generated by the stimulated emission is highly *monochromatic*, *coherent*, and directional. In the stimulated emission, one incident photon generates two photons; one is the incident photon itself, and the other is an emitted photon due to the stimulated emission. As a result, the incident light is amplified by the stimulated emission. The absorption depicted in Fig. 2.4 (c) is a process that the electron transits from a lower energy state to a higher one by absorbing energy from the incident light. Because this transition is induced by the incident light, it is sometimes called *induced absorption*. It should be noted that spontaneous absorption does not exist; this will be explained in Chapter 8.

When a light is incident on a material, the stimulated emission and the absorption simultaneously take place. In *thermal equilibrium*, there are more electrons in a lower energy state than in a higher one, because the lower energy state is more stable than the higher one. Therefore, in thermal equilibrium, only the absorption is observed when a light is incident on a material. In order to obtain a net optical gain, we have to make the number of electrons in a higher energy state larger than in the lower one. This condition is referred to as *inverted population*, or *population inversion*, because the electron

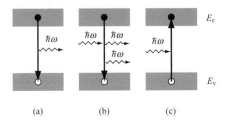

Fig. 2.4. Radiation and absorption: (a) spontaneous emission, (b) stimulated emission, and (c) absorption

population is inverted compared with that in thermal equilibrium. In semiconductors, the population inversion is obtained only in the vicinity of the band edges by excitation of the electrons through optical pumping or electric current injection. The population inversion generates many electrons at the bottom of the conduction band and many holes at the top of the valence bands.

The laser oscillators use fractions of the spontaneous emission as the optical input and amplify the fractions by the stimulated emission under population inversion. Once the optical gains exceed the optical losses in the laser oscillators, *laser oscillations* take place. The term *laser* is an acronym for "light amplification by stimulated emission of radiation" and is used as a noun in a *laser oscillator*. Note that an emitted light from a laser oscillator is not a laser but a *laser light* or a *laser beam*. As a back formation from laser, "lase" is used as a verb meaning "to emit coherent light."

Among semiconductor light emitting devices, LEDs use spontaneous emission and are applied to remote-control transmitters, switch lights, brake lights, displays, and traffic signals. In contrast, LDs are oscillators of lights using stimulated emission and are used as light sources for lightwave communications, compact discs (CDs), magneto-optical discs (MOs), digital video discs (DVDs), laser beam printers, laser pointers, bar-code readers, and so on.

2.4 Optical Gains

2.4.1 Lasers

Let us consider the relationships between *laser oscillators* and *optical gains*. First, we review a general oscillator, as shown in Fig. 2.5. An oscillator has a *gain* and amplifies an input. Also, the oscillator returns a fraction of an output to an input port through a *feedback* loop. The returned one is repetitively amplified, and oscillation starts when a net gain exceeds an internal loss of the oscillator.

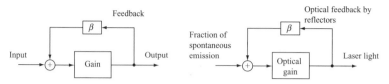

Fig. 2.5. Oscillator   **Fig. 2.6.** Laser oscillator

Figure 2.6 shows a laser oscillator where a fraction of the spontaneous emission is used as the input and the *optical gain* is produced by the stimulated emission. To feed back light, *optical resonators* or *optical cavities*, which are composed of *reflectors* or *mirrors*, are adopted. Due to this configuration, characteristics of lasers (for brevity, *lasers* are used for laser oscillators here) are affected by the optical gains and optical resonators. Many lasers use optical resonators, but nitrogen lasers, whose optical gains are very large, can start laser oscillations even without optical resonators. It should be noted that all semiconductor lasers use optical resonators.

As explained earlier, a fraction of the spontaneous emission is used as the input of a laser. Note that all the spontaneously emitted lights cannot be used as the input, because they have different wavelengths, phases, and propagation directions. Among these lights, only the lights, which have wavelengths within the optical gain spectrum and satisfy the resonance conditions of the optical resonators, can be the sources of laser oscillations. Other spontaneous emissions, which do not satisfy the resonance conditions of optical resonators, are readily emitted outward without obtaining sufficient optical gains for laser oscillations. The lights amplified by the stimulated emission have the same wavelength, phase, and propagation direction as those of the input lights. Therefore, the laser lights are highly monochromatic, bright, coherent, and directional.

2.4.2 Optical Gains

(a) Carrier Distribution

We derive equations for the optical gains of semiconductor lasers. We suppose that many electrons are excited to the conduction band by optical pumping or electrical current injection. In this condition, the *carrier distribution* is in *nonthermal equilibrium*, and there are many electrons in the conduction band and many holes in the valence band. As a result, one *Fermi level* E_F cannot describe the distribution functions of the electrons and holes. In this case, it is useful to determine the distribution functions by assuming that the electrons in the conduction band and the holes in the valence band are separately governed by *Fermi-Dirac distribution*. For this purpose, we introduce *quasi-Fermi levels* E_{Fc} and E_{Fv} defined as

$$n = N_c \exp\left(-\frac{E_c - E_{Fc}}{k_B T}\right),$$

$$p = N_v \exp\left(-\frac{E_{Fv} - E_v}{k_B T}\right),$$

$$N_c = 2 \left(\frac{2\pi m_e^* k_B T}{h^2}\right)^{3/2},$$

$$N_v = 2 \left(\frac{2\pi m_h^* k_B T}{h^2}\right)^{3/2}.$$

(2.1)

Here, n and p are the electron concentration and the hole concentration, respectively; E_c and E_v are the energy of the bottom of the conduction band and that of the top of the valence band, respectively; $k_B = 1.3807 \times 10^{-23}\,\mathrm{J\,K^{-1}}$ is *Boltzmann constant*; T is an absolute temperature; N_c and N_v are the *effective density of states* for the electrons and that for the holes, respectively; m_e^* and m_h^* are the effective mass of the electrons and that of the holes, respectively; and h is Planck's constant.

From (2.1), these quasi-Fermi levels E_{Fc} and E_{Fv} are written as

$$E_{Fc} = E_c + k_B T \ln\left(\frac{n}{N_c}\right),$$

$$E_{Fv} = E_v - k_B T \ln\left(\frac{p}{N_v}\right).$$

(2.2)

We express a distribution function for the electrons in the valence band with the energy E_1 as f_1 and that for the electrons in the conduction band with the energy E_2 as f_2. Using E_{Fc} and E_{Fv}, we can express f_1 and f_2 as

$$f_1 = \frac{1}{\exp\left[(E_1 - E_{Fv})/(k_B T)\right] + 1},$$

$$f_2 = \frac{1}{\exp\left[(E_2 - E_{Fc})/(k_B T)\right] + 1}.$$

(2.3)

It should be noted that the distribution function for the holes in the valence band is given by $[1 - f_1]$.

(b) Optical Transition Rates

As shown in Fig. 2.7, we assume that a light, which has a photon energy of $E_{21} = E_2 - E_1$ and a photon density of $n_{ph}(E_{21})$, interacts with a direct-transition semiconductor. Under this assumption, we calculate the *transition rates* for the stimulated emission, the absorption, and the spontaneous emission. In Fig. 2.7, E_c is the energy for the bottom of the conduction band, and E_v is the energy for the top of the valence band. Note that this figure shows the energy bands for a certain value of $\boldsymbol{k}$, and the horizontal line has no physical meaning.

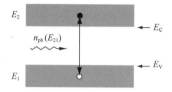

Fig. 2.7. Schematic model for radiation and absorption

(i) Stimulated Emission Rate

The *stimulated emission rate per unit volume* r_{21}(stim) is given by a product of

the transition probability per unit time from E_2 to $E_1 : B_{21}$,
the electron concentration in a state with the energy $E_2 : n_2$,
the hole concentration in a state with the energy $E_1 : p_1$,
the photon density : $n_{\mathrm{ph}}(E_{21})$.

The concentration n_2 of the electron, which occupies a state with the energy E_2 in the conduction band, is expressed as

$$n_2 = \rho_{\mathrm{c}}(E_2 - E_{\mathrm{c}})f_2, \tag{2.4}$$

where $\rho_{\mathrm{c}}(E_2 - E_{\mathrm{c}})$ is the density of states, which is a function of $E_2 - E_{\mathrm{c}}$, and f_2 is the distribution function in (2.3).

The concentration p_1 of the hole, which occupies a state with the energy E_1 in the valence band, is written as

$$p_1 = \rho_{\mathrm{v}}(E_{\mathrm{v}} - E_1)[1 - f_1], \tag{2.5}$$

where $\rho_{\mathrm{v}}(E_{\mathrm{v}} - E_1)$ is the density of states, which is a function of $E_{\mathrm{v}} - E_1$, and f_1 is the distribution function in (2.3).

Therefore, the stimulated emission rate $r_{21}(stim)$ is obtained as

$$
\begin{aligned}
r_{21}(\mathrm{stim}) &= B_{21}n_2p_1n_{\mathrm{ph}}(E_{21}) \\
&= B_{21}n_{\mathrm{ph}}(E_{21})\rho_{\mathrm{c}}(E_2 - E_{\mathrm{c}})\rho_{\mathrm{v}}(E_{\mathrm{v}} - E_1)f_2[1 - f_1].
\end{aligned} \tag{2.6}
$$

(ii) Absorption Rate

The *absorption rate per unit volume* r_{12}(abs) is given by a product of

the transition probability per unit time from E_1 to $E_2 : B_{12}$,
the concentration of an empty state with the energy $E_2 : p_2$,
the electron concentration in a state with the energy $E_1 : n_1$,
the photon density : $n_{\mathrm{ph}}(E_{21})$.

The concentration p_2 of an empty state, which is not occupied by the electrons with the energy E_2 in the conduction band, is expressed as

$$p_2 = \rho_c(E_2 - E_c)[1 - f_2]. \tag{2.7}$$

The concentration n_1 of the electron, which occupies a state with the energy E_1 in the valence band, is written as

$$n_1 = \rho_v(E_v - E_1)f_1. \tag{2.8}$$

As a result, the absorption rate $r_{12}(\text{abs})$ is obtained as

$$\begin{aligned} r_{12}(\text{abs}) &= B_{12}p_2 n_1 n_{\text{ph}}(E_{21}) \\ &= B_{12}n_{\text{ph}}(E_{21})\rho_c(E_2 - E_c)\rho_v(E_v - E_1)f_1[1 - f_2]. \end{aligned} \tag{2.9}$$

(iii) Spontaneous Emission Rate

The *spontaneous emission rate per unit volume* $r_{21}(\text{spon})$ is independent of the incident photon density and given by a product of

> the transition probability per unit time from E_2 to E_1 : A_{21},
> the electron concentration in a state with the energy E_2 : n_2,
> the hole concentration in a state with the energy E_1 : p_1.

Using (2.4) and (2.5), the spontaneous emission rate $r_{21}(\text{spon})$ is obtained as

$$\begin{aligned} r_{21}(\text{spon}) &= A_{21}n_2 p_1 \\ &= A_{21}\rho_c(E_2 - E_c)\rho_v(E_v - E_1)f_2[1 - f_1]. \end{aligned} \tag{2.10}$$

(c) Optical Transition in Thermal Equilibrium

In thermal equilibrium, the carrier distributions are described by one Fermi level E_F, and the radiations balance the absorption. In other words, a sum of the stimulated emission rate and the spontaneous emission rate is equal to the absorption rate. Using the optical transition rates for the stimulated emission, the spontaneous emission, and the absorption, we obtain

$$\begin{aligned} r_{21}(\text{stim}) + r_{21}(\text{spon}) &= r_{12}(\text{abs}), \\ E_{F1} = E_{F2} &= E_F. \end{aligned} \tag{2.11}$$

Substituting (2.6), (2.9), and (2.10) into (2.11), we have

$$n_{\text{ph}}(E_{21}) = \frac{A_{21}}{B_{12}\exp[E_{21}/(k_B T)] - B_{21}}, \tag{2.12}$$

while the *blackbody radiation theory* gives

$$n_{\mathrm{ph}}(E_{21}) = \frac{8\pi n_{\mathrm{r}}^{3} E_{21}^{2}}{h^{3}c^{3}\exp[E_{21}/(k_{\mathrm{B}}T)] - h^{3}c^{3}}. \tag{2.13}$$

Here, n_{r} is the effective refractive index of a material, h is Planck's constant, and $c = 2.99792458 \times 10^{8}\,\mathrm{m\,s^{-1}}$ is the speed of light in a vacuum.

Comparison of (2.12) and (2.13) results in

$$A_{21} = Z(E_{21})B, \quad B_{21} = B_{12} = B,$$
$$Z(E_{21}) = \frac{8\pi n_{\mathrm{r}}^{3} E_{21}^{2}}{h^{3}c^{3}}, \tag{2.14}$$

which is known as *Einstein's relation*, and A_{21} is called *Einstein's A coefficient*; and B is called *Einstein's B coefficient*. It should be noted that $B_{21} = B_{12} = B$, and A_{21} is proportional to B.

We consider the physical meaning of $Z(E_{21})$ in (2.14) in the following. We suppose that a light field E_{L} is written as

$$E_{\mathrm{L}} = A_{0}\exp[\mathrm{i}\,(\omega t - \boldsymbol{k} \cdot \boldsymbol{r})], \tag{2.15}$$

where A_{0} is a complex amplitude, ω is an angular frequency, t is a time, $\boldsymbol{k}$ is a wave vector, and $\boldsymbol{r}$ is a position vector. We also suppose that the periodic boundary condition is satisfied on the surfaces of a cube with a side length L. Because the electric fields at $x = 0$ and $x = L$ are the same, an eigenmode is given by

$$\exp(0) = \exp(\mathrm{i}\,k_{x}L) = 1, \quad \therefore \ k_{x} = \frac{2\pi}{L}n_{x}, \tag{2.16}$$

where k_{x} is an x-component of the wave vector $\boldsymbol{k}$ and n_{x} is an integer. Similarly, we obtain

$$k_{y} = \frac{2\pi}{L}n_{y}, \quad k_{z} = \frac{2\pi}{L}n_{z}, \tag{2.17}$$

where n_{y} and n_{z} are integers. Hence, $k^{2} = k_{x}^{2} + k_{y}^{2} + k_{z}^{2}$ is expressed as

$$k^{2} = \left(\frac{2\pi}{L}\right)^{2}(n_{x}^{2} + n_{y}^{2} + n_{z}^{2}). \tag{2.18}$$

Using the effective refractive index n_{r} of the material, the angular frequency ω is given by

$$\omega = \frac{kc}{n_{\mathrm{r}}}. \tag{2.19}$$

Therefore, the energy $E = \hbar\omega$ is written as

$$E = \hbar\omega = \frac{2\pi\hbar c}{n_{\mathrm{r}}L}(n_{x}^{2} + n_{y}^{2} + n_{z}^{2})^{1/2} = \frac{hc}{n_{\mathrm{r}}L}(n_{x}^{2} + n_{y}^{2} + n_{z}^{2})^{1/2}, \tag{2.20}$$

which takes discrete values, because n_x, n_y, and n_z are integers. However, when L is much larger than the wavelength of the light, $n_x{}^2 + n_y{}^2 + n_z{}^2$ becomes a huge number, and many modes exist near each other. When we consider a sphere with a radius of $R = (n_x{}^2 + n_y{}^2 + n_z{}^2)^{1/2}$, the number of combinations (n_x, n_y, n_z) is equal to a volume of the sphere $4\pi R^3/3$. For each combination (n_x, n_y, n_z), there exist two modes corresponding to their polarizations, which are perpendicular to each other. As a result, the number of modes whose energy is placed between 0 and E is given by

$$2 \times \frac{4\pi}{3} \left(\frac{n_\mathrm{r} E L}{hc}\right)^3 = \frac{8\pi n_\mathrm{r}{}^3 E^3}{3h^3 c^3} L^3. \tag{2.21}$$

The number of modes with their energies between E and $E + \mathrm{d}E$ is obtained as a derivative of (2.21) with respect to E, which is expressed as

$$\frac{8\pi n_\mathrm{r}{}^3 E^2}{h^3 c^3} L^3 \, \mathrm{d}E. \tag{2.22}$$

Because a volume of the cube is L^3, dividing (2.22) by L^3 gives the number of modes per unit volume, that is, the *mode density* $m(E) \, \mathrm{d}E$ as

$$m(E) \, \mathrm{d}E = \frac{8\pi n_\mathrm{r}{}^3 E^2}{h^3 c^3} \, \mathrm{d}E. \tag{2.23}$$

From (2.23) and the third equation in (2.14), it is found that we have $m(E_{21}) = Z(E_{21})$. This relation suggests that the spontaneous emission takes place for all the modes with their energy placed between E and $E+\mathrm{d}E$ while the stimulated emission and absorption occur only for the mode corresponding to the incident light.

Note that the derivation of A and B assumes that both $r_{21}(\text{stim})$ and $r_{12}(\text{abs})$ are proportional to the photon density $n_{\text{ph}}(E_{21})$. If we assume that both $r_{21}(\text{stim})$ and $r_{12}(\text{abs})$ are proportional to the energy density of the photons as Einstein did [7], other formulas for A and B will be obtained.

Note: Definitions of the Mode Density

In the preceding discussion, we explained the mode density $m(E) \, \mathrm{d}E$ with the energies placed between E and $E + \mathrm{d}E$. Here, we will derive mode densities $m(\omega) \, \mathrm{d}\omega$ with the angular frequencies of a light between ω and $\omega + \mathrm{d}\omega$, and $m(\nu) \, \mathrm{d}\nu$ with the light frequencies between ν and $\nu+\mathrm{d}\nu$. Using the (effective) refractive index of the material n_r and the speed of light in a vacuum c, we have

$$k = \frac{n_\mathrm{r}\omega}{c} = \frac{2\pi n_\mathrm{r}\nu}{c}. \tag{2.24}$$

Substituting (2.24) into (2.18) and then calculating a volume $4\pi R^3/3$ of a sphere with the radius $R = (n_x{}^2 + n_y{}^2 + n_z{}^2)^{1/2}$ results in

$$2 \times \frac{4\pi}{3}R^3 = \frac{n_{\rm r}{}^3\omega^3}{3\pi^2c^3}L^3 = \frac{8\pi n_{\rm r}{}^3\nu^3}{3c^3}L^3, \tag{2.25}$$

where a factor 2 corresponds to the directions of the polarizations. As a result, dividing (2.25) by L^3 and then differentiating with respect to ω or ν leads to

$$m(\omega)\,\mathrm{d}\omega = \frac{n_{\rm r}{}^3\omega^2}{\pi^2c^3}\,\mathrm{d}\omega, \quad m(\nu)\,\mathrm{d}\nu = \frac{8\pi n_{\rm r}{}^3\nu^2}{c^3}\,\mathrm{d}\nu. \tag{2.26}$$

(d) Net Stimulated Emission

Let us consider the excited conditions where many free carriers exist. In these nonthermal equilibrium conditions, radiations do not balance with absorption anymore. When a light is incident on a material, the stimulated emission and the absorption simultaneously take place. Therefore, the net stimulated emission rate $r^0(\mathrm{stim})$ is given by

$$
\begin{aligned}
r^0(\mathrm{stim}) &= r_{21}(\mathrm{stim}) - r_{12}(\mathrm{abs}) \\
&= Bn_{\rm ph}(E_{21})\rho_{\rm c}(E_2 - E_{\rm c})\rho_{\rm v}(E_{\rm v} - E_1)[f_2 - f_1] \\
&= \frac{A_{21}}{Z(E_{21})}n_{\rm ph}(E_{21})\rho_{\rm c}(E_2 - E_{\rm c})\rho_{\rm v}(E_{\rm v} - E_1)[f_2 - f_1]. \tag{2.27}
\end{aligned}
$$

From (2.27), to obtain the *net stimulated emission* $r^0(\mathrm{stim}) > 0$, we need

$$f_2 > f_1, \tag{2.28}$$

which indicates the population inversion in the semiconductors. With the help of (2.3), (2.28) reduces to

$$E_{\rm Fc} - E_{\rm Fv} > E_2 - E_1 = E_{21}, \tag{2.29}$$

which is known as the *Bernard-Duraffourg relation*. In semiconductor lasers, a typical carrier concentration for $r^0(\mathrm{stim}) > 0$ is of the order of $10^{18}\,\mathrm{cm}^{-3}$.

(e) Net Stimulated Emission Rate and Optical Gains

The optical power gain coefficient per unit length g is defined as

$$\frac{\mathrm{d}I}{\mathrm{d}z} = gI, \tag{2.30}$$

where I is the light intensity per unit area and z is a coordinate for a propagation direction of the light. Because the light intensity is proportional to the square of the electric field, the amplitude gain coefficient $g_{\rm E}$, which is the gain coefficient for the electric field, is given by $g_{\rm E} = g/2$. From the definition, I is expressed as

$$I = v E_{21} n_{\text{ph}}(E_{21}) = v \hbar \omega n_{\text{ph}}(E_{21}),$$

$$v = \frac{d\omega}{dk} = \frac{c}{n_{\text{r}}}, \tag{2.31}$$

$$\frac{dI}{dt} = v E_{21} \frac{dn_{\text{ph}}(E_{21})}{dt} = v E_{21} r^0(\text{stim}),$$

where v is a group velocity of the light in the material, $E_{21} = \hbar\omega$ is the photon energy, and $n_{\text{ph}}(E_{21})$ is the photon density.

Here we consider a relationship between the derivatives of I with respect to a position and a time. Because the position z is a function of the time t, we have

$$\frac{dI}{dz} = \frac{dt}{dz}\frac{dI}{dt} = \left(\frac{dz}{dt}\right)^{-1}\frac{dI}{dt} = \frac{1}{v}\frac{dI}{dt}, \tag{2.32}$$

$$\frac{dI}{dz} = gI = g v E_{21} n_{\text{ph}}(E_{21}).$$

From (2.31) and (2.32), we have the relation

$$r^0(\text{stim}) = v g n_{\text{ph}}(E_{21}). \tag{2.33}$$

As a result, it is found that a large $r^0(\text{stim})$ leads to a large g. Using (2.27), (2.31), and (2.33), we can also express g as

$$g = \frac{r^0(\text{stim})}{n_{\text{ph}}(E_{21})}\frac{n_{\text{r}}}{c} = \frac{n_{\text{r}}}{c}B\rho_{\text{c}}(E_2 - E_{\text{c}})\rho_{\text{v}}(E_{\text{v}} - E_1)[f_2 - f_1]. \tag{2.34}$$

From the *time-dependent quantum mechanical perturbation theory* (see Appendix C), Einstein's B coefficient in (2.14), which is the transition rate due to the stimulated emission for a photon in a free space, is given by

$$B = \frac{e^2 h}{2m^2 \varepsilon_0 n_{\text{r}}^2 E_{21}}\langle 1|\boldsymbol{p}|2\rangle^2. \tag{2.35}$$

Here, e is the elementary charge; h is Planck's constant; m is the electron mass; ε_0 is permittivity in a vacuum; n_{r} is the refractive index of the material; E_{21} is the photon energy; $\boldsymbol{p}$ is the momentum operator; and $|1\rangle$ and $|2\rangle$ are the wave functions of the valence band and the conduction band in a steady state, respectively.

Substituting (2.35) into (2.14), we can write Einstein's A coefficient as

$$A_{21} = \frac{4\pi e^2 n_{\text{r}} E_{21}}{m^2 \varepsilon_0 h^2 c^3}\langle 1|\boldsymbol{p}|2\rangle^2. \tag{2.36}$$

Using the *dipole moment* μ, (2.35) and (2.36) can be rewritten as follows: The momentum operator $\boldsymbol{p}$ is expressed as

$$p = m \frac{\mathrm{d}r}{\mathrm{d}t}. \tag{2.37}$$

As a result, we obtain

$$\frac{1}{m}\langle 1|p|2\rangle = \frac{\mathrm{d}}{\mathrm{d}t}\langle 1|r|2\rangle = \mathrm{i}\,\omega\langle 1|r|2\rangle, \tag{2.38}$$

where $r \propto \mathrm{e}^{\mathrm{i}\omega t}$ was assumed. With the help of (2.38), Einstein's A and B coefficients are written as

$$B = \frac{\pi e^2 \omega}{\varepsilon_0 n_r{}^2}\langle 1|r|2\rangle^2 = \frac{\pi\omega}{\varepsilon_0 n_r{}^2}\mu^2,$$

$$A_{21} = \frac{2n_r\omega^3}{\varepsilon_0 hc^3}\mu^2, \tag{2.39}$$

$$\mu^2 = \langle 1|er|2\rangle^2.$$

Note that many textbooks on quantum electronics use (2.39) as Einstein's A coefficient. Expressions on Einstein's B coefficient depend on the definitions of the stimulated emission rate whether it is proportional to the photon density or to the energy density. In the preceding explanations, Einstein's B coefficient is defined to be in proportion to the photon density. Therefore, the quantum mechanical transition rate for the stimulated emission is equal to Einstein's B coefficient.

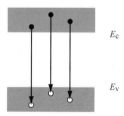

Fig. 2.8. Transition with a constant photon energy

In semiconductors, due to the energy band structures, transitions take place between various energy states, as shown in Fig. 2.8. If we put $E_{21} = E$ and $E_2 - E_c = E''$, the electron energy in the valence band E' for the allowed transition is given by $E' = E'' - E$. Therefore, by integrating (2.27) with respect to E'', the net stimulated emission rate $r^0(\text{stim})$ in the semiconductors is obtained as

$$r^0(\text{stim}) = \frac{e^2 h\, n_{\mathrm{ph}}(E)}{2m^2\varepsilon_0 n_r{}^2 E}\int_0^\infty \langle 1|p|2\rangle^2 \rho_c(E'')\rho_v(E')[f_2(E'') - f_1(E')]\,\mathrm{d}E'', \tag{2.40}$$

where (2.35) was used. Similarly, the optical power gain coefficient $g(E)$ is given by

$$g(E) = \frac{e^2 h}{2m^2 \varepsilon_0 n_r c E} \int_0^\infty \langle 1|\boldsymbol{p}|2\rangle^2 \rho_c(E'')\rho_v(E')[f_2(E'') - f_1(E')]\,\mathrm{d}E''.$$

(2.41)

From (2.10) and (2.36), the spontaneous emission rate $r_{21}(\text{spon})$ is expressed as

$$r_{21}(\text{spon}) = \frac{4\pi e^2 n_r E}{m^2 \varepsilon_0 h^2 c^3} \int_0^\infty \langle 1|\boldsymbol{p}|2\rangle^2 \rho_c(E'')\rho_v(E')f_2(E'')[1 - f_1(E')]\,\mathrm{d}E''.$$

(2.42)

When the excitation is weak, the absorption is observed, and the net absorption rate $r^0(abs)$ is written as

$$r^0(\text{abs}) = r_{12}(\text{abs}) - r_{21}(\text{stim}) = -r^0(\text{stim}).$$

(2.43)

Therefore, the optical power absorption coefficient $\alpha(E)$ and the optical power gain coefficient $g(E)$ are related as

$$\alpha(E) = -g(E).$$

(2.44)

Figure 2.9 shows the calculated optical power gain coefficient $g(E)$ in (2.41) as a function of $E - E_g$ for the bulk structures with the bandgap energy of $E_g = 0.8\,\text{eV}$. With an increase in the carrier concentration n, the gain peak shifts toward a higher energy (shorter wavelength). This shift of the gain peak is caused by the *band filling effect* [8], in which the electrons in the conduction band and the holes in the valence band occupy each band from the band edges. Because higher energy states are more dense than lower energy states, the optical gain in higher energy states is larger than that in lower energy ones.

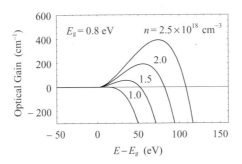

Fig. 2.9. Optical power gain coefficient

Here, let us consider the optical power gain coefficient $g(E)$ from other viewpoints. In (2.41), $\rho_c(E'')\rho_v(E')$ is considered to be the density of state pairs related to the optical transitions. The density of state pairs is also expressed using the law of momentum conservation ($\boldsymbol{k}$-*selection rule*) and the law of energy conservation ($E_{21} = E_2 - E_1$). Under the $\boldsymbol{k}$-selection rule, optical transitions take place only for $\boldsymbol{k} = \boldsymbol{k}_1 = \boldsymbol{k}_2$. Therefore, we can define the density of state pairs or the reduced density of states $\rho_{\mathrm{red}}(E_{21})$ as

$$\rho_{\mathrm{red}}(E_{21})\,\mathrm{d}E_{21} \equiv \rho_{\mathrm{v}}(E_1)\,\mathrm{d}E_1 = \rho_{\mathrm{c}}(E_2)\,\mathrm{d}E_2, \tag{2.45}$$

where

$$\mathrm{d}E_{21} = \mathrm{d}E_1 + \mathrm{d}E_2. \tag{2.46}$$

From (2.45) and (2.46), we have

$$\rho_{\mathrm{red}}(E_{21}) = \left[\frac{1}{\rho_{\mathrm{v}}(E_1)} + \frac{1}{\rho_{\mathrm{c}}(E_2)}\right]^{-1}. \tag{2.47}$$

If the conduction and valence bands are parabolic, the energies E_1 and E_2 are written as

$$E_1 = E_{\mathrm{v}} - \frac{\hbar^2 k^2}{2m_{\mathrm{h}}{}^*}, \tag{2.48}$$

$$E_2 = E_{\mathrm{c}} + \frac{\hbar^2 k^2}{2m_{\mathrm{e}}{}^*}, \tag{2.49}$$

where $m_{\mathrm{h}}{}^*$ and $m_{\mathrm{e}}{}^*$ are the effective masses of the holes and the electrons, respectively. As a result, E_1 and E_2 are rewritten as

$$E_1 = E_{\mathrm{v}} - \frac{m_{\mathrm{e}}{}^*}{m_{\mathrm{e}}{}^* + m_{\mathrm{h}}{}^*}\,(E_{21} - E_{\mathrm{g}}), \tag{2.50}$$

$$E_2 = E_{\mathrm{c}} + \frac{m_{\mathrm{h}}{}^*}{m_{\mathrm{e}}{}^* + m_{\mathrm{h}}{}^*}\,(E_{21} - E_{\mathrm{g}}). \tag{2.51}$$

Also, once the initial state of the optical transitions is given, the final state is determined by E_{21} and $\boldsymbol{k}$. Hence, the number of state pairs per volume associated with the optical transitions for the photon energies between E_{21} and $E_{21} + \mathrm{d}E_{21}$ is given by

$$\rho_{\mathrm{red}}(E_{21})\,[f_2(E_{21}) - f_1(E_{21})]\,\mathrm{d}E_{21}. \tag{2.52}$$

We assume that the energy states related to optical transitions have an energy width of $\hbar/\tau_{\mathrm{in}}$, and the spectral shape of Lorentzian such as

$$L(E_{21}) = \frac{1}{\pi}\,\frac{\hbar/\tau_{\mathrm{in}}}{(E_{21} - E)^2 + (\hbar/\tau_{\mathrm{in}})^2}, \tag{2.53}$$

where τ_{in} is the relaxation time due to electron scatterings and transitions. In this case, the optical power gain coefficient $g(E)$ is expressed as

$$g(E) = \frac{e^2 h}{2m^2 \varepsilon_0 n_r c E} \int_0^\infty \langle 1|\boldsymbol{p}|2\rangle^2 \rho_{\text{red}}(E_{21}) \left[f_2(E_{21}) - f_1(E_{21})\right] L(E_{21}) \, dE_{21},$$

$$(2.54)$$

which is plotted in Fig. 2.10 for the bulk structures with $\tau_{\text{in}} = 10^{-13}$ s. It is found that relaxation is equivalent to band tailing, which contributes to optical transitions in $E < E_{\text{g}}$.

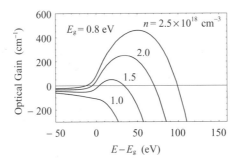

Fig. 2.10. Optical power gain coefficient under the $\boldsymbol{k}$-selection rule

Because f_2 and $[1 - f_1]$ are always positive, spontaneously emitted lights distribute in a higher energy than the gain peak. Figure 2.11 schematically shows spectra for the stimulated emission (laser light) and the spontaneously emitted lights; its horizontal line indicates photon energy. Note that the photon energies increase due to the band filling effect with enhancement of the excitation, irrespective of spontaneous emission or stimulated emission.

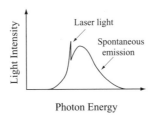

Fig. 2.11. Stimulated emission and spontaneous emission

3 Optical Waveguides

3.1 Introduction

An optical beam propagating in a *free space* expands in the beam width with an increase in propagation distance. The beam divergence angle θ in radians is expressed as

$$\theta \simeq \frac{\lambda}{d}, \tag{3.1}$$

where λ is a light wavelength and d is a beam diameter. For example, if a laser beam with $\lambda = 1\,\mu\mathrm{m}$ and $d = 1\,\mathrm{mm}$ propagates by $100\,\mathrm{m}$, its beam width expands to about $10\,\mathrm{cm}$. In contrast, the *optical waveguide* confines lights to itself during propagation of the lights [9–13].

In typical bulk (double heterostructure) semiconductor lasers, the active layer thickness is $0.1\,\mu\mathrm{m}$, the beam width is $0.3\,\mu\mathrm{m}$, and the cavity length is $300\,\mu\mathrm{m}$. If there are no optical waveguides, this beam width expands up to $500\,\mu\mathrm{m}$ by a single-way propagation, where λ is assumed to be $0.5\,\mu\mathrm{m}$. In the semiconductor lasers, only the *active layer* has the optical gain, and it is extremely inefficient to amplify a beam with the width of $500\,\mu\mathrm{m}$ by the active layer with the thickness of only $0.1\,\mu\mathrm{m}$. Therefore, to amplify lights efficiently, optical waveguides are indispensable for semiconductor lasers. It should be noted that *optical fibers* are also representative optical waveguides.

Figure 3.1 shows cross-sectional views of the optical waveguides. According to the operating principles, optical waveguides are divided into *index guiding* waveguides and *gain guiding* ones. In index guiding waveguides, a light is confined to a high refractive index region, which is surrounded by low refractive index regions. In Fig. 3.1 (a), only if the *refractive indexes* n_f, n_c, and n_s satisfy $n_\mathrm{f} > n_\mathrm{c}, n_\mathrm{s}$, a light is confined to a region with n_f. The gain guiding waveguides use a property that only a gain region amplifies a light, and a light seems to propagate in the gain region. In Fig. 3.1 (b), a light is confined to a region with the *optical gain* g_f, which is larger than g_c and g_s.

To consider optical gain, we introduce the *complex refractive index*

$$\tilde{n} = n_\mathrm{r} - \mathrm{i}\,\kappa, \tag{3.2}$$

where n_r is a so-called refractive index and κ is the *extinction coefficient*. Here, we express the electric field E of a light as

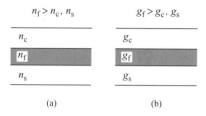

Fig. 3.1. Cross sections of optical waveguides: (a) index guiding and (b) gain guiding

$$E = E_0 \exp[\mathrm{i}\,(\omega t - kz)], \tag{3.3}$$

where E_0 is an amplitude, ω is an angular frequency of the light, t is a time, k is a wave number, and z is a coordinate for a propagation direction of the light. Using the complex refractive index $\tilde{n}$ in (3.2), the angular frequency ω, and the speed of light in a vacuum c, we can write the wave number k as

$$k = \frac{\tilde{n}\omega}{c} = \frac{n_{\mathrm{r}} - \mathrm{i}\,\kappa}{c}\omega. \tag{3.4}$$

Substituting (3.4) into (3.3), we have

$$E = E_0 \exp\left[\mathrm{i}\left(\omega t - \frac{n_{\mathrm{r}}\omega}{c}z\right)\right] \cdot \exp\left(-\frac{\omega\kappa}{c}z\right), \tag{3.5}$$

where $\kappa > 0$ shows the optical loss and $\kappa < 0$ indicates the optical gain. As a result, the *optical amplitude gain coefficient* g_{E} and the extinction coefficient κ are related as

$$g_{\mathrm{E}} = -\frac{\omega}{c}\kappa. \tag{3.6}$$

Note that the optical power gain coefficient g is equal to $2g_{\mathrm{E}}$, because the light intensity is proportional to E^2.

From the viewpoint of shape, the optical waveguides are classified into a *two-dimensional optical waveguide*, or a *planar optical waveguide*, and a *three-dimensional optical waveguide*, or a *strip optical waveguide*. The two-dimensional waveguide has a plane, which is much larger than a wavelength of light and confines a light one-dimensionally. Only the size of its confinement direction is on the order of a light wavelength or less. The three-dimensional waveguide confines a light two-dimensionally, and the sizes along these two confinement directions are on the order of a light wavelength or less. The remaining direction is the propagation direction of a light.

3.2 Two-Dimensional Optical Waveguides

3.2.1 Propagation Modes

Let us consider a two-dimensional waveguide where the *guiding layer* is sandwiched by the *cladding layer* and the *substrate*, as shown in Fig. 3.2. The refractive indexes of the guiding layer, the cladding layer, and the substrate are n_f, n_c, and n_s, respectively. To confine a light in the guiding layer, we need $n_f > n_s \geq n_c$, and $n_f - n_s$ is usually on the order of 10^{-3} to 10^{-1}.

Cladding layer n_c

Guiding layer n_f

Substrate n_s

Fig. 3.2. Cross section of a two-dimensional optical waveguide

Figure 3.3 shows propagation directions of a light when the light enters from the substrate to the cladding layer through the guiding layer. From *Snell's law*, the angles θ_s, θ_f, and θ_c, which are formed by the interface normals and the directions of the light, are related to the refractive indexes n_s, n_f, and n_c as

$$n_s \sin \theta_s = n_f \sin \theta_f = n_c \sin \theta_c. \tag{3.7}$$

Fig. 3.3. Snell's law

When θ_s is equal to $\pi/2$, a light cannot propagate in the substrate, and the *total reflection* takes place at the interface of the substrate and the guiding layer. In this case, the power reflectivity (reflectance) is 100%. When θ_c is equal to $\pi/2$, the total reflection takes place at the interface of the cladding layer and the guiding layer. The minimum value of θ_f to obtain the total reflection is called the *critical angle*. According to $\theta_s = \pi/2$, and $\theta_c = \pi/2$, we have two critical angles θ_{fs} and θ_{fc}, which are expressed as

$$\theta_{fs} = \sin^{-1}\left(\frac{n_s}{n_f}\right), \quad \theta_{fc} = \sin^{-1}\left(\frac{n_c}{n_f}\right). \tag{3.8}$$

Here, $\theta_{fs} \geq \theta_{fc}$ is satisfied under the assumption of $n_f > n_s \geq n_c$.

Corresponding to a value of θ_f, there are three *propagation modes*, as shown in Fig. 3.4.

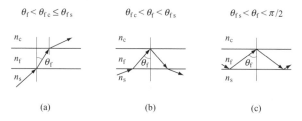

Fig. 3.4. Propagation modes: (a) radiation mode, (b) substrate radiation mode, and (c) guided mode

(a) $\theta_f < \theta_{fc} \leq \theta_{fs}$: Radiation Mode

The light is not confined to the optical waveguide at all.

(b) $\theta_{fc} < \theta_f < \theta_{fs}$: Substrate Radiation Mode

The total reflection occurs at the interface of the cladding layer and the guiding layer, while the refraction takes place at the interface of the guiding layer and the substrate. As a result, there exists a light propagating in the substrate.

(c) $\theta_{fs} < \theta_f < \pi/2$: Guided Mode

At both the interfaces, the total reflections occur. A light is completely confined to the optical waveguide.

Among (a), (b), and (c), the guided mode in (c) is the most important for semiconductor lasers and photonic integrated circuits. In (a) and (b), the guided mode does not exist, and such a condition is referred to as the *cutoff*. Therefore, in the following, we will examine the guided mode in detail.

3.2.2 Guided Mode

(a) Total Reflection

The *total reflection* is the reflection with $|r|^2 = 1$, where r is the amplitude reflectivity. It should be noted that r depends on the *angle of incidence* and the *direction of polarization*.

Here, we consider a *plane wave*, and we define the *plane of incidence* as a plane on which all the directions of *incident light*, *reflected light*, and *refracted light* coexist. The *transverse electric* (TE) mode is a *linearly polarized light* whose *electric field* E is normal to the plane of incidence. The *transverse magnetic* (TM) mode is a linearly polarized light whose *magnetic field* H is

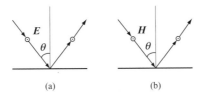

Fig. 3.5. (a) TE mode and (b) TM mode

normal to the plane of incidence. Figure 3.5 shows the TE mode and the TM mode, where the plane of incidence is placed on the face of the page and only the incident and reflected lights are illustrated.

From the *Fresnel formulas*, the amplitude reflectivities are given by

$$r_{\text{TE,c}} = \frac{n_f \cos\theta_f - \sqrt{n_c^2 - n_f^2 \sin^2\theta_f}}{n_f \cos\theta_f + \sqrt{n_c^2 - n_f^2 \sin^2\theta_f}},$$

$$(3.9)$$

$$r_{\text{TE,s}} = \frac{n_f \cos\theta_f - \sqrt{n_s^2 - n_f^2 \sin^2\theta_f}}{n_f \cos\theta_f + \sqrt{n_s^2 - n_f^2 \sin^2\theta_f}},$$

$$r_{\text{TM,c}} = \frac{n_c^2 \cos\theta_f - n_f \sqrt{n_c^2 - n_f^2 \sin^2\theta_f}}{n_c^2 \cos\theta_f + n_f \sqrt{n_c^2 - n_f^2 \sin^2\theta_f}},$$

$$(3.10)$$

$$r_{\text{TM,s}} = \frac{n_s^2 \cos\theta_f - n_f \sqrt{n_s^2 - n_f^2 \sin^2\theta_f}}{n_s^2 \cos\theta_f + n_f \sqrt{n_s^2 - n_f^2 \sin^2\theta_f}}.$$

Here, subscripts TE and TM show the TE mode and the TM mode, respectively; subscripts c and s correspond to the reflection at the interface of the cladding layer and the guiding layer and that at the interface of the substrate and the guiding layer, respectively.

When the total reflections take place, $|r|^2 = 1$ is satisfied, and the amplitude reflectivities in (3.9) and (3.10) take complex numbers. As a result, we can write the amplitude reflectivity r as

$$r = \exp(\mathrm{i}\,2\phi), \qquad (3.11)$$

where ϕ is the *phase shift* added to the reflected wave at the interface. In other words, due to the total reflection, the phase of the reflected wave is shifted at the interface of the reflection. When we rewrite (3.9) and (3.10) in the form of (3.11), the phase shifts ϕ are expressed as

$$\tan\phi_{\mathrm{TE,c}} = \frac{\sqrt{n_{\mathrm{f}}{}^2 \sin^2\theta_{\mathrm{f}} - n_{\mathrm{c}}{}^2}}{n_{\mathrm{f}}\cos\theta_{\mathrm{f}}},$$

$$\tag{3.12}$$

$$\tan\phi_{\mathrm{TE,s}} = \frac{\sqrt{n_{\mathrm{f}}{}^2 \sin^2\theta_{\mathrm{f}} - n_{\mathrm{s}}{}^2}}{n_{\mathrm{f}}\cos\theta_{\mathrm{f}}},$$

$$\tan\phi_{\mathrm{TM,c}} = \frac{n_{\mathrm{f}}\sqrt{n_{\mathrm{f}}{}^2 \sin^2\theta_{\mathrm{f}} - n_{\mathrm{c}}{}^2}}{n_{\mathrm{c}}{}^2\cos\theta_{\mathrm{f}}},$$

$$\tag{3.13}$$

$$\tan\phi_{\mathrm{TM,s}} = \frac{n_{\mathrm{f}}\sqrt{n_{\mathrm{f}}{}^2 \sin^2\theta_{\mathrm{f}} - n_{\mathrm{s}}{}^2}}{n_{\mathrm{s}}{}^2\cos\theta_{\mathrm{f}}},$$

where the subscripts are the same as in (3.9) and (3.10).

(b) Guiding Condition

We consider a monochromatic coherent plane wave, which propagates in an optical waveguide as shown in Fig. 3.6. Here, the x-axis is normal to the layer interfaces, the y-axis is parallel to the layer interfaces, and the z-axis is along the propagation direction of the light. The thickness of the guiding layer is h.

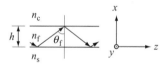

Fig. 3.6. Coordinate system for a guided mode

Neglecting a time-dependent factor, we can express the propagating electric field E as

$$E = E_0 \exp[-\mathrm{i}\,k_0 n_{\mathrm{f}}(\pm x\cos\theta_{\mathrm{f}} + z\sin\theta_{\mathrm{f}})], \tag{3.14}$$

where $k_0 = \omega/c$ is a wave number in a vacuum, and a sign in front of x corresponds to the propagation direction of the light ($+$: toward positive x and $-$: toward negative x). The *propagation constant* β and the *phase velocity* v_{p} along the z-axis are defined as

$$\beta \equiv k_0 n_{\mathrm{f}} \sin\theta_{\mathrm{f}} \equiv \frac{\omega}{v_{\mathrm{p}}}, \tag{3.15}$$

as shown in Fig. 3.7.

$$\beta = k_0\, n_f \sin\theta_f$$

Fig. 3.7. Propagation constant

If we see a light in a reference system, which moves toward the positive z-axis with a phase velocity v_p, we observe that a light is multireflected along the x-axis. Changes in the phase during a roundtrip of the light along the x-axis are given by

$$
\begin{aligned}
&x = 0 \rightarrow h \ \text{(the first half)} && : k_0 n_f h \cos\theta_f, \\
&x = h \ \quad \text{(at the interface of the reflection)} && : -2\phi_c, \\
&x = h \rightarrow 0 \ \text{(the second half)} && : k_0 n_f h \cos\theta_f, \\
&x = 0 \ \quad \text{(at the interface of the reflection)} && : -2\phi_s.
\end{aligned}
$$

Because the phase of the wave front shifts after a roundtrip, the lightwave intensity is modified by interference. Therefore, to obtain a propagating lightwave without decay, a total phase change in the roundtrip must satisfy

$$2k_0 n_f h \cos\theta_f - 2\phi_c - 2\phi_s = 2m\pi \quad (m = 0, 1, 2, \cdots), \tag{3.16}$$

where m is the *mode number*. Equation (3.16) shows a resonance condition normal to the z-axis, which is called the *transverse resonance condition*. From (3.16), it is found that the angle of incidence θ_f for the guided mode takes discrete values.

In summary, the guiding condition is characterized by simultaneously satisfying the transverse resonance condition (3.16) and the total reflection condition $\theta_{fs} < \theta_f < \pi/2$.

(c) Goos-Hänchen Shift

When the total reflection takes place, a phase of a lightwave shifts by -2ϕ at the interface of the reflection. Due to this phase shift, an optical path changes at the reflection interfaces, as shown in Fig. 3.8. This shift in the optical path Z is called the *Goos-Hänchen shift*, which is given by

$$Z = \frac{d\phi}{d\beta}, \tag{3.17}$$

where β is the propagation constant along the z-axis.

At each reflection interface, a lightwave propagates along the interface and has a component with a decay constant $1/X$ along the x-axis where X is the *penetration depth*. This exponentially decaying wave is designated the *evanescent wave*, and the energy does not flow along the x-axis.

With the help of (3.12) and (3.13), the Goos-Hänchen shifts and the penetration depths are expressed as

$$k_0 Z_{\mathrm{TE,c}} = (n_{\mathrm{f}}^2 \sin^2 \theta_{\mathrm{f}} - n_{\mathrm{c}}^2)^{-1/2} \cdot \tan \theta_{\mathrm{f}},$$
$$k_0 Z_{\mathrm{TE,s}} = (n_{\mathrm{f}}^2 \sin^2 \theta_{\mathrm{f}} - n_{\mathrm{s}}^2)^{-1/2} \cdot \tan \theta_{\mathrm{f}}, \tag{3.18}$$
$$X_{\mathrm{TE,c}} = \frac{Z_{\mathrm{TE,c}}}{\tan \theta_{\mathrm{f}}}, \quad X_{\mathrm{TE,s}} = \frac{Z_{\mathrm{TE,s}}}{\tan \theta_{\mathrm{f}}},$$

$$k_0 Z_{\mathrm{TM,c}} = (n_{\mathrm{f}}^2 \sin^2 \theta_{\mathrm{f}} - n_{\mathrm{c}}^2)^{-1/2} \cdot \tan \theta_{\mathrm{f}} \cdot \left(\frac{N^2}{n_{\mathrm{c}}^2} + \frac{N^2}{n_{\mathrm{f}}^2} - 1 \right)^{-1},$$

$$k_0 Z_{\mathrm{TM,s}} = (n_{\mathrm{f}}^2 \sin^2 \theta_{\mathrm{f}} - n_{\mathrm{s}}^2)^{-1/2} \cdot \tan \theta_{\mathrm{f}} \cdot \left(\frac{N^2}{n_{\mathrm{s}}^2} + \frac{N^2}{n_{\mathrm{f}}^2} - 1 \right)^{-1}, \tag{3.19}$$

$$X_{\mathrm{TM,c}} = \frac{Z_{\mathrm{TM,c}}}{\tan \theta_{\mathrm{f}}}, \quad X_{\mathrm{TM,s}} = \frac{Z_{\mathrm{TM,s}}}{\tan \theta_{\mathrm{f}}}.$$

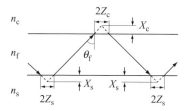

Fig. 3.8. Goos-Hänchen shift

(d) Effective Refractive Index

Using a wave number $k_0 = \omega/c$ in a vacuum and the propagation constant β along the z-axis, we define the *effective refractive index* N as

$$N \equiv \frac{\beta}{k_0} = n_{\mathrm{f}} \sin \theta_{\mathrm{f}}. \tag{3.20}$$

This effective refractive index N is considered to be a refractive index of a material for a plane wave propagating along the z-axis. Because the semiconductor lasers have optical waveguides within themselves, the effective refractive index N is important to determine the resonance conditions. Note that the value of N for the guided modes satisfies

$$n_{\mathrm{s}} < N < n_{\mathrm{f}}, \tag{3.21}$$

where $n_{\mathrm{f}} > n_{\mathrm{s}}$ was used.

(e) Optical Confinement Factor

We define the *optical confinement factor* Γ as a ratio of the intensity of the light existing in the relevant layer to the total light intensity.

Because the light intensity distributes as shown in Fig. 3.9, the optical confinement factor Γ_f for the guiding layer (hatched area) is given by

$$\Gamma_f = \frac{\displaystyle\int_0^h |E(x)|^2 \mathrm{d}x}{\displaystyle\int_{-\infty}^{\infty} |E(x)|^2 \mathrm{d}x}. \tag{3.22}$$

Similarly, the optical confinement factors Γ_c for the cladding layer and Γ_s for the substrate are written as

$$\Gamma_c = \frac{\displaystyle\int_h^{\infty} |E(x)|^2 \mathrm{d}x}{\displaystyle\int_{-\infty}^{\infty} |E(x)|^2 \mathrm{d}x}, \quad \Gamma_s = \frac{\displaystyle\int_{-\infty}^0 |E(x)|^2 \mathrm{d}x}{\displaystyle\int_{-\infty}^{\infty} |E(x)|^2 \mathrm{d}x}. \tag{3.23}$$

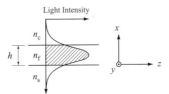

Fig. 3.9. Distribution of light intensity

The optical confinement factor Γ is important to design the optical losses or the optical gains in the optical waveguides. Using Γ, we can approximately express the effective refractive index N as

$$N \approx \Gamma_s n_s + \Gamma_f n_f + \Gamma_c n_c. \tag{3.24}$$

(f) Normalized Expressions for Eigenvalue Equation

The equation for the transverse resonance condition (3.16) is called an *eigenvalue equation*. By solving (3.16) numerically, we can examine propagation characteristics of the guided modes. Furthermore, normalizing (3.16) leads to *dispersion curves*, which are common to all step-like two-dimensional optical waveguides. To obtain a normalized expression of the eigenvalue equation, we define the *normalized frequency*, or the *normalized waveguide thickness* V, and the *normalized waveguide refractive index* b as

$$V = k_0 h \sqrt{n_f{}^2 - n_s{}^2}, \tag{3.25}$$

$$b = \frac{N^2 - n_s{}^2}{n_f{}^2 - n_s{}^2}. \tag{3.26}$$

In addition, we introduce the *asymmetry measure* a as

$$a = \frac{n_s{}^2 - n_c{}^2}{n_f{}^2 - n_s{}^2}. \tag{3.27}$$

Substituting (3.12), (3.13), and (3.25)–(3.27) into (3.16), we obtain the normalized eigenvalue equation as

$$V\sqrt{1-b} = m\pi + \tan^{-1}\chi_s\sqrt{\frac{b}{1-b}} + \tan^{-1}\chi_c\sqrt{\frac{a+b}{1-b}}, \tag{3.28}$$

where

$$\chi_i = \begin{cases} 1 & : \text{TE mode} \\ (n_f/n_i)^2 & : \text{TM mode } (i = s, c). \end{cases} \tag{3.29}$$

It should be noted that the guided modes for the gain guiding waveguides can be calculated from (3.16) or (3.28), if we replace the refractive index n_i ($i = s, f, c$) with the complex refractive index $\tilde{n}_i$.

Figure 3.10 shows the normalized dispersion curves for the guided TE modes. If the waveguide parameters, such as the refractive indexes and the layer thickness, are given, we can design the optical waveguides using Fig. 3.10. For the design of the optical waveguides, the *cutoff condition*, in which the guided modes do not exist, is important. When the angle of incidence θ_f is equal to the critical angle θ_{fs}, the lightwave is not confined to the guiding layer anymore, and a fraction of the optical power is emitted to the substrate. In this case, we have $N = n_s$, which results in $b = 0$ from (3.26). Hence, from (3.28), a normalized frequency for the cutoff condition V_m is given by

$$V_m = m\pi + \tan^{-1}\chi_c\sqrt{a}. \tag{3.30}$$

For $V_m < V < V_{m+1}$, the guided modes from the zeroth order to the mth order exist as shown in Fig. 3.10, and a lower-order mode has a larger b for a common V value. Therefore, from (3.26), a lower-order mode has a larger effective refractive index, and the fundamental mode ($m = 0$) has the largest N among the guided modes.

Figure 3.11 shows the electric fields $E(x)$ for fundamental ($m = 0$), first-order ($m = 1$), and second-order ($m = 2$) TE modes, in which the mth-order TE mode is indicated by TE_m. See Appendix D for the relationship between the eigenvalue equations and the light fields.

For symmetric optical waveguides with $n_s = n_c$, by using V and b, the optical confinement factor Γ_f is also expressed as

$$\Gamma_f = \frac{V\sqrt{b} + 2b}{V\sqrt{b} + 2}. \tag{3.31}$$

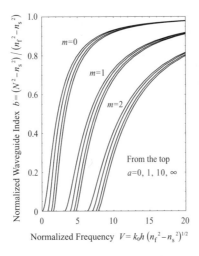

Fig. 3.10. Normalized dispersion curves for guided TE modes

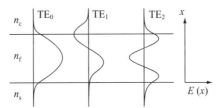

Fig. 3.11. Distributions of electric fields

(g) Coupling of Light into Optical Waveguides

When the total reflections occur at the interfaces, we cannot couple a light into the guiding layer from the cladding layer or the substrate. Therefore, we introduce the light from a *facet* of the optical waveguide, as shown in Fig. 3.12. It should be noted that the angle of incidence in the guiding layer θ_f must satisfy the guiding conditions.

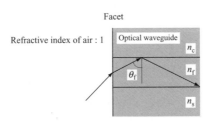

Fig. 3.12. Coupling of a light into an optical waveguide

In semiconductor lasers, lights are generated in the active layers, and we do not have to couple a light from outside. However, when semiconductor lasers are used as optical amplifiers by biasing below threshold, or they are controlled by an incident light such as in *injection locking*, lights are introduced from the facets of semiconductor lasers.

3.3 Three-Dimensional Optical Waveguides

As shown in Fig. 3.13, we select a coordinate system in which the propagation direction of the light is the z-axis, the layer interfaces are normal to the x-axis, and the layer planes are parallel to the y-axis.

The two-dimensional optical waveguides can confine lights only along the x-axis, not along the y-axis. In contrast, the *three-dimensional optical waveguides* confine lights along both the x- and y-axes. When these three-dimensional optical waveguides are adopted in semiconductor lasers, the guided modes are efficiently amplified, which results in low-threshold, high-efficiency laser operations. However, analysis of the three-dimensional optical waveguides is more complicated than that of two-dimensional ones, and we cannot obtain exact analytical solutions. As a result, approximate analytical methods or numerical analyses are used to design the three-dimensional optical waveguides. As approximate analytical methods, we explain the *effective refractive index method* and *Marcatili's method*. These approximate methods can be applied to the guided modes, only when an *aspect ratio $a/h > 1$* where h is the guiding layer thickness and a is the waveguide width, and the guiding condition is far from the cutoff conditions. If these two conditions are not satisfied, accuracy of the approximations degrades.

Fig. 3.13. Coordinates for a three-dimensional optical waveguide

3.3.1 Effective Refractive Index Method

Figure 3.14 shows a *ridge* optical waveguide where an upper figure is a top view and a lower one is a cross-sectional view. Here, a is the waveguide width, h is the film thickness of the core region, and f is the film thickness of the surrounding regions.

At first, we treat the core region and the surrounding regions separately, and we regard each region as the two-dimensional optical waveguide. Therefore, the normalized frequency V and the normalized waveguide refractive index b for each region are defined as

$$V_\mathrm{h} = k_0 h \sqrt{n_\mathrm{f}^2 - n_\mathrm{s}^2}, \quad V_\mathrm{f} = k_0 f \sqrt{n_\mathrm{f}^2 - n_\mathrm{s}^2}, \tag{3.32}$$

$$b_\mathrm{h} = \frac{N_\mathrm{h}^2 - n_\mathrm{s}^2}{n_\mathrm{f}^2 - n_\mathrm{s}^2}, \quad b_\mathrm{f} = \frac{N_\mathrm{f}^2 - n_\mathrm{s}^2}{n_\mathrm{f}^2 - n_\mathrm{s}^2}, \tag{3.33}$$

where subscripts h and f correspond to the core region and the surrounding regions, respectively; and N_h and N_f are the effective refractive indexes for each region.

Secondly, we consider that the guiding layer with the refractive index N_h and thickness a is sandwiched by the layers with the refractive index N_f, as shown in the upper figure of Fig. 3.14. As a result, the normalized frequency V_y of the three-dimensional optical waveguide is defined as

$$V_y = k_0 a \sqrt{N_\mathrm{h}^2 - N_\mathrm{f}^2}. \tag{3.34}$$

If we express the effective refractive index of the three-dimensional optical waveguide as N_s, the normalized waveguide refractive index b_s of the three-dimensional optical waveguide is given by

$$b_\mathrm{s} = \frac{N_\mathrm{s}^2 - N_\mathrm{f}^2}{N_\mathrm{h}^2 - N_\mathrm{f}^2}. \tag{3.35}$$

Substituting V_y and b_s into (3.28) leads to the dispersion curves for the three-dimensional optical waveguide.

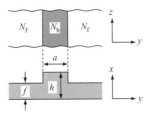

Fig. 3.14. Effective refractive index method

3.3.2 Marcatili's Method

Figure 3.15 shows a cross-sectional view of the three-dimensional optical waveguide, which is seen toward the positive direction of the z-axis. If most of the guided modes are confined to region I, light field amplitudes drastically decay with an increase in distance from the interfaces. Therefore, the light intensities distributed in the shaded areas in Fig. 3.15 are neglected, which is referred to as Marcatili's method [14]. Under this assumption, light fields are obtained by solving wave equations.

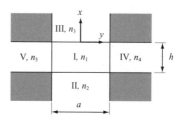

Fig. 3.15. Marcatili's method

4 Optical Resonators

4.1 Introduction

As shown in Fig. 4.1, the lasers use a fraction of spontaneously emitted lights as the input and amplify the fraction by the stimulated emission. To feed back the light, the *optical resonators*, or the *optical cavities*, which consist of reflectors, are adopted. The resonance conditions of the optical resonators determine lasing conditions such as a threshold and an oscillation wavelength.

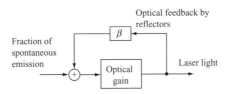

Fig. 4.1. Laser

In this chapter, we focus on the optical resonators that are used in the semiconductor lasers and explain the *Fabry-Perot cavity*, the *distributed feedback* (DFB) cavity, and the *distributed Bragg reflector* (DBR), which is a component of a resonator.

The optical resonators are divided into three groups whose constituent components are *mirrors*, *diffraction gratings*, and hybrid of mirrors and diffraction gratings.

Figure 4.2 shows the optical resonators comprising the mirrors: the *Fabry-Perot cavity* and the *ring cavity*. The Fabry-Perot cavity consists of two parallel mirrors in which lights experience repetitive roundtrips. The ring cavity, in contrast, has three or more mirrors, in which lights propagate clockwise or counterclockwise. These two optical cavities are also widely used in gas, solid, and dye lasers as well as in semiconductor lasers. One of the mirrors is often replaced by the diffraction grating in order to select an oscillation mode. Note that a *cleaved facet* is used as a mirror in semiconductor lasers.

Figure 4.3 shows the optical resonators, which use diffraction gratings: the *DFB* cavity and the *DBR*. The DFB cavity has an active layer, which generates a light, and the optical gain, in its corrugated region, and functions

(a)   (b)

Fig. 4.2. Optical resonators with reflectors: (a) Fabry-Perot cavity and (b) ring cavity

as the optical resonator by itself. In contrast, the DBR does not have an active layer in its corrugated region and functions as a reflector, not a resonator. Therefore, the DBR is combined with other DBRs or the cleaved facets to form optical resonators.

(a)   (b)

Fig. 4.3. (a) DFB and (b) DBR

4.2 Fabry-Perot Cavity

The *Fabry-Perot laser* has an active material inside the Fabry-Perot cavity, as shown in Fig. 4.4 (a). In the semiconductor lasers, cleaved facets are used as mirrors, as illustrated in Fig. 4.4 (b), because lights are reflected at the interface of the air and the semiconductor due to a difference in refractive indexes. Note that the cleaved facets are very flat with the order of atomic layers, and they are much smoother than light wavelengths. These cleaved facets are often coated with *dielectric films* to control the reflectivities or protect the facets from degradation.

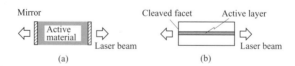

(a)   (b)

Fig. 4.4. Fabry-Perot lasers: (a) laser and (b) LD

Let us consider *transmission characteristics* of the Fabry-Perot cavity. We assume that the amplitude reflectivities of the two mirrors are r_1 and r_2; the amplitude transmissivities of the two mirrors are t_1 and t_2, as shown in

Fig. 4.5. As explained in Chapter 3, these reflectivities and transmissivities depend on the angle of incidence and the polarization of the light. Here, we suppose that the angle of incidence θ_0 is small enough, and we regard r_1, r_2, t_1, and t_2 as constant.

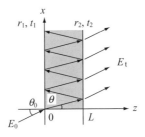

Fig. 4.5. Analytical model for the Fabry-Perot cavity

If we neglect a time-dependent factor, we can write the incident electric field E_i through the surface at $z = 0$ as

$$E_i = t_1 E_0 \exp\left[-\frac{i\,n_{rt}\omega}{c}(x\sin\theta + z\cos\theta)\right] \times \exp[g_E(x\sin\theta + z\cos\theta)]. \quad (4.1)$$

Here, E_0 is an amplitude of the electric field of the incident light, n_{rt} is a real part of the complex refractive index of the material placed between the two mirrors, ω is an angular frequency of the light, c is the speed of light in a vacuum, and g_E is the optical amplitude gain coefficient defined in (3.6).

The transmitted electric field E_t through the Fabry-Perot cavity is expressed as

$$
\begin{aligned}
E_t &= t_1 t_2 E_0 \exp\left[-\frac{i\,n_{rt}\omega}{c}(x\sin\theta + L\cos\theta) + \frac{g_E L}{\cos\theta}\right] \\
&\quad \times \left\{1 + r_1 r_2 \exp\left(-i\delta + \frac{2g_E L}{\cos\theta}\right) + (r_1 r_2)^2 \exp\left[2\left(-i\delta + \frac{2g_E L}{\cos\theta}\right)\right]\right. \\
&\qquad \left. + \cdots\right\} \\
&= \frac{t_1 t_2 E_0 \exp\left[-\dfrac{i\,n_{rt}\omega}{c}(x\sin\theta + L\cos\theta) + \dfrac{g_E L}{\cos\theta}\right]}{1 - r_1 r_2 \exp\left(-i\delta + \dfrac{2g_E L}{\cos\theta}\right)},
\end{aligned}
\quad (4.2)
$$

where

$$\delta = \frac{2n_{rt}\omega L\cos\theta}{c} \quad (4.3)$$

was introduced.

The incident light intensity I_0 and the transmitted light intensity I_t are related to the electric fields E_0 and E_t as

$$I_0 \propto E_0^* E_0, \quad I_t \propto E_t^* E_t. \tag{4.4}$$

Therefore, with the help of (4.2), we obtain

$$I_t = \frac{(t_1 t_2)^2 G_s}{1 + (r_1 r_2)^2 \cdot G_s^{\,2} - 2r_1 r_2 G_s \cos \delta} I_0,$$

$$G_s = \exp\left(\frac{2g_E L}{\cos \theta}\right). \tag{4.5}$$

Because the power reflectivities R_1, R_2 and the power transmissivities T_1, T_2 are given by

$$R_1 = r_1^* r_1, \quad R_2 = r_2^* r_2, \quad T_1 = t_1^* t_1, \quad T_2 = t_2^* t_2, \tag{4.6}$$

we can rewrite (4.5) as

$$I_t = \frac{T_1 T_2 G_s}{1 + G_s^{\,2} R_1 R_2 - 2G_s \sqrt{R_1 R_2} \cos \delta} I_0. \tag{4.7}$$

The angle θ is given by $\theta = \pi/2 - \theta_f$, where θ_f was defined in Fig. 3.6. Hence, we have $\cos \theta = \sin \theta_f$ and (4.3) reduces to

$$\delta = \frac{2n_r \omega L}{c} = 2n_r k_0 L, \quad k_0 = \frac{\omega}{c}, \tag{4.8}$$

where the effective refractive index $N = n_{rt} \sin \theta_f \equiv n_r$ was used. When the angle of incidence θ_0 is small, as in a fundamental mode, (4.7) results in

$$I_t = \frac{T_1 T_2 G_{s0}}{(1 - G_{s0}\sqrt{R_1 R_2})^2 + 4G_{s0}\sqrt{R_1 R_2}\sin^2(n_r k_0 L)} I_0,$$

$$G_{s0} = \exp(2g_E L) = \exp(gL), \tag{4.9}$$

where (4.8) was used and g is the optical power gain coefficient. Figure 4.6 shows a plot of (4.9) where $G_{s0} = 1$, and $R_1 = R_2 = R$, $T_1 = T_2 = T = 0.98 - R$ by assuming that the optical power loss at each mirror is 2%. The transmissivities have maximum values at $n_r k_0 L = n\pi$ (n: a positive integer) and minimum values at $(n + 1/2)\pi$. With an increase in R, the transmission spectra narrow and the transmissivities decrease. If the optical power loss at the mirror is null, we have $T = 1 - R$, which results in a maximum power transmissivity of 1 (100%) irrespective of R.

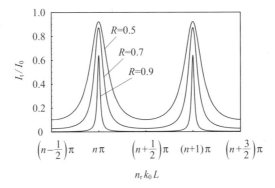

Fig. 4.6. Transmission characteristics for the Fabry-Perot cavity

4.2.1 Resonance Condition

When the *resonance condition* is satisfied, the power transmissivity has a peak. From (4.9), the resonance condition is given by

$$n_r k_0 L = \frac{n_r \omega L}{c} = \frac{n_{rt} \omega L}{c} \cos\theta = n\pi, \tag{4.10}$$

where n is a positive integer. From this equation, it is found that the effective refractive index n_r is useful to express the resonance condition. At *normal incidence* with $\theta_0 = \theta = 0$, the resonance condition is written as

$$\frac{n_{rt} \omega L}{c} = n_{rt} k_0 L = n\pi. \tag{4.11}$$

Using a wavelength in a vacuum λ_0, (4.11) reduces to

$$L = n \frac{\lambda_0}{2n_{rt}}. \tag{4.12}$$

Because λ_0/n_{rt} is a wavelength in a material, a product of a positive integer and a half-wavelength in a material is equal to the cavity length L at the resonance condition.

4.2.2 Free Spectral Range

The Fabry-Perot cavity is used as a *spectrometer*, because its transmissivity depends on a light wavelength. However, two lights whose $n_r k_0 L$ values are different by $n\pi$ (n: a positive integer) cannot be resolved by the Fabry-Perot cavity, because there is a common transmissivity for the two lights. When $(n-1/2)\pi < n_r k_0 L < (n+1/2)\pi$ is satisfied, a light is not confused by other lights. This region of $n_r k_0 L$ is called the *free spectral range*, because the light is resolved free from other lights. The free spectral range in an angular frequency ω_{FSR} is given by

$$\omega_{\text{FSR}} = \frac{c}{n_r L}\pi. \tag{4.13}$$

Using the free spectral range λ_{FSR} in a wavelength, ω_{FSR} is written as

$$\omega_{\text{FSR}} = 2\pi c \left(\frac{1}{\lambda_0} - \frac{1}{\lambda_0 + \lambda_{\text{FSR}}} \right). \tag{4.14}$$

As a result, we have

$$\lambda_{\text{FSR}} \simeq \frac{\lambda_0{}^2}{2 n_r L} = \frac{\lambda_0{}^2}{2\pi c}\omega_{\text{FSR}}. \tag{4.15}$$

4.2.3 Spectral Linewidth

The *half width at half maximum* (HWHM) is a difference between the wavelength or the frequency for the maximum transmissivity and that for the half of the maximum transmissivity. The *full width at half maximum* (FWHM), which is twice as large as the HWHM, is a difference between the two wavelengths or frequencies for the half of the maximum transmissivity.

Let us calculate the *spectral linewidth* using (4.9). When the transmissivity takes a maximum value, the resonance condition (4.11) is satisfied. As a result, the denominator in (4.9) reduces to

$$(1 - G_{s0}\sqrt{R_1 R_2})^2. \tag{4.16}$$

When the transmissivity is half-maximum, the denominator in (4.9) is twice (4.16), and we have

$$(1 - G_{s0}\sqrt{R_1 R_2})^2 + 4 G_{s0}\sqrt{R_1 R_2}\sin^2(n_r k_0 L) = 2(1 - G_{s0}\sqrt{R_1 R_2})^2, \tag{4.17}$$

which results in

$$\sin(n_r k_0 L) = \pm \frac{1}{2\sqrt[4]{G_{s0}{}^2 R_1 R_2}}(1 - G_{s0}\sqrt{R_1 R_2}). \tag{4.18}$$

Expressing $k_0 = k_m$ on resonance and $k_0 = k_m \pm \Delta k_0$ at half-maximum and then substituting $k_0 = k_m \pm \Delta k_0$ into (4.18) leads to

$$\sin(n_r \Delta k_0 L) = \frac{1}{2\sqrt[4]{G_{s0}{}^2 R_1 R_2}}(1 - G_{s0}\sqrt{R_1 R_2}). \tag{4.19}$$

For the spectra in Fig. 4.6, $n_r \Delta k_0 L \ll 1$ is satisfied. Hence, (4.19) reduces to

$$\Delta k_0 = \frac{1 - G_{s0}\sqrt{R_1 R_2}}{2n_r L \sqrt[4]{G_{s0}^2 R_1 R_2}}. \tag{4.20}$$

Using $k_0 = \omega/c$, we can express the HWHM in an angular frequency $\Delta\omega_H$ as

$$\Delta\omega_H = c\Delta k_0 = \frac{c(1 - G_{s0}\sqrt{R_1 R_2})}{2n_r L \sqrt[4]{G_{s0}^2 R_1 R_2}}. \tag{4.21}$$

Therefore, the FWHM $\Delta\omega_F$, which is twice the HWHM, is given by

$$\Delta\omega_F = 2\Delta\omega_H = \frac{c(1 - G_{s0}\sqrt{R_1 R_2})}{n_r L \sqrt[4]{G_{s0}^2 R_1 R_2}}. \tag{4.22}$$

Using a wavelength in a vacuum, we can write the FWHM in a wavelength $\Delta\lambda_F$ as

$$\Delta\lambda_F \simeq \frac{\lambda_0^2}{2\pi c}\Delta\omega_F = \frac{\lambda_0^2(1 - G_{s0}\sqrt{R_1 R_2})}{2\pi n_r L \sqrt[4]{G_{s0}^2 R_1 R_2}}. \tag{4.23}$$

4.2.4 Finesse

When we measure a wavelength or a frequency of light, an index of resolution is given by the *finesse*. The finesse F is defined as a ratio of the free spectral range to the FWHM, which is given by

$$F = \frac{\omega_{FSR}}{\Delta\omega_F} = \frac{\lambda_{FSR}}{\Delta\lambda_F} = \frac{\pi\sqrt[4]{G_{s0}^2 R_1 R_2}}{1 - G_{s0}\sqrt{R_1 R_2}}, \tag{4.24}$$

where (4.13), (4.15), (4.22), and (4.23) were used. With an increase in F, the resolution of a wavelength or a frequency is improved.

4.2.5 Electric Field Inside Fabry-Perot Cavity

Earlier, we considered a relationship between the incident light and the transmitted light in the Fabry-Perot cavity. In this section, we study the light inside the Fabry-Perot cavity. When the amplitude reflectivities are $r_1 = r_2 = r$ and the amplitude transmissivities are $t_1 = t_2 = t$, the electric field E inside the Fabry-Perot cavity is written as

$$\begin{aligned} E &= tE_0 \exp(k_{eff}\, x \sin\theta)\left\{\exp(k_{eff}\, z \cos\theta) + r\exp[k_{eff}(2L - z)\cos\theta]\right\} \\ &\quad \times \left[1 + r^2 \exp(2k_{eff} L \cos\theta) + r^4 \exp(4k_{eff} L \cos\theta) + \cdots\right] \\[2mm] &= \frac{tE_0 \exp(k_{eff}\, x \sin\theta)\left\{\exp(k_{eff}\, z \cos\theta) + r\exp[k_{eff}(2L - z)\cos\theta]\right\}}{1 - r^2 \exp(2k_{eff} L \cos\theta)}, \end{aligned} \tag{4.25}$$

where

$$k_{\mathrm{eff}} = -\frac{\mathrm{i}\, n_{\mathrm{rt}}\omega}{c} + g_{\mathrm{E}}. \tag{4.26}$$

In (4.25), $\exp(k_{\mathrm{eff}}\, z\cos\theta)$ exhibits a forward running wave toward a positive direction along the z-axis, and $\exp[k_{\mathrm{eff}}\,(2L - z)\cos\theta]$ expresses a backward running wave after reflection at a plane with $z = L$. If we introduce $a = r\exp(2k_{\mathrm{eff}}\,L\cos\theta)$, (4.25) reduces to

$$E = tE_0 \exp(k_{\mathrm{eff}}\, x\sin\theta)\frac{(1 - a)\exp(k_{\mathrm{eff}}\, z\cos\theta) + 2a\cosh(k_{\mathrm{eff}}\, z\cos\theta)}{1 - r^2\exp(2k_{\mathrm{eff}}\,L\cos\theta)}.$$
$$\tag{4.27}$$

In (4.27), the first term in the numerator represents a forward *running wave*, and the second term shows a *standing wave*. From (4.25)–(4.27), the light intensity inside the Fabry-Perot cavity takes a maximum value when

$$\frac{n_{\mathrm{rt}}\omega L}{c}\cos\theta = \frac{n_{\mathrm{r}}\omega L}{c} = n_{\mathrm{r}}k_0 L = n\pi. \tag{4.28}$$

Here, n is a positive integer, which shows the number of *nodes* existing between $z = 0$ and $z = L$ for a standing wave. Note that (4.28) is the same as the resonance condition (4.10), which indicates that both the transmissivity and the internal light intensity of the Fabry-Perot cavity have the largest values at the resonance condition.

4.3 DFB and DBR

The *DFB* cavity and the *DBR* are the optical waveguides, which have diffraction gratings in them. They feed back lights by spatially modulating the complex refractive indexes of the optical waveguides. In the Fabry-Perot cavity, the reflection points of lights are only the facets. In contrast, in the DFB cavity and the DBR, reflection points of lights are distributed in the optical waveguide as shown in Fig. 4.7. The difference in the DFB cavity and the DBR is that the former has the optical gain in the corrugated region, and the latter does not. As described earlier, the DFB cavity functions as the optical resonator, and the DBR forms the optical resonator with other DBRs or cleaved facets.

4.3.1 Coupled Wave Theory [15]

Because only the difference in the DFB cavity and the DBR is whether there is optical gain, their characteristics can be analyzed by a common method. In order to treat the optical gains or losses, we consider an optical waveguide with a complex dielectric constant $\tilde{\varepsilon}$.

Mirror   Mirror

(a)   (b)

Fig. 4.7. Feedback points of lights: (a) Fabry-Perot cavity and (b) DFB cavity and DBR

When a lightwave is assumed to propagate along the z-axis, a propagation constant k in the optical waveguide is given by

$$k^2 = \omega^2 \mu \tilde{\varepsilon} = \omega^2 \mu (\varepsilon_r - \mathrm{i}\,\varepsilon_i) \simeq k_0^2 n_r(z)^2 \left[1 + \mathrm{i}\,\frac{2\alpha(z)}{k_0 n_r(z)} \right],$$

(4.29)

$$\tilde{\varepsilon} = \varepsilon_r - \mathrm{i}\,\varepsilon_i = \varepsilon_0 \left[n_r(z) + \mathrm{i}\,\frac{\alpha(z)}{k_0} \right]^2,$$

where ω is an angular frequency of the light; μ is *permeability* of a material; $k_0 = \omega/c$ is a wave number in a vacuum; $\varepsilon_0 = 8.854 \times 10^{-12}\,\mathrm{F/m}$ is *permittivity in a vacuum*; $n_r(z)$ is a real part of a complex refractive index; and $\alpha(z) = g_E$ is an optical amplitude gain coefficient. For the usual optical materials, μ is almost equal to *permeability in a vacuum* $\mu_0 = 4\pi \times 10^{-7}\,\mathrm{H/m}$. Because of $|\alpha(z)| \ll k_0$, a second-order term of $\alpha(z)$ was neglected in (4.29). Figure 4.8 shows a schematic cross section of the diffraction gratings. The effective refractive index of the optical waveguide is periodically modulated with a pitch of Λ by the corrugations formed at the interface between the two layers with the refractive indexes n_A and n_B ($n_A \neq n_B$).

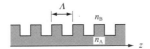

Fig. 4.8. Diffraction grating

Here, we assume that $n_r(z)$ and $\alpha(z)$ are sinusoidal functions of z, which are written as

$$n_r(z) = n_{r0} + n_{r1} \cos(2\beta_0 z + \Omega),$$
$$\alpha(z) = \alpha_0 + \alpha_1 \cos(2\beta_0 z + \Omega),$$

(4.30)

where Ω is a phase at $z = 0$ and β_0 is related to the grating pitch Λ as

$$\beta_0 = \frac{\pi}{\Lambda}.$$

(4.31)

Under the assumption of $n_{r1} \ll n_{r0}$ and $\alpha_1 \ll \alpha_0$, substituting (4.30) into (4.29) leads to

$$k(z)^2 = k_0{}^2 n_{r0}{}^2 + i\, 2k_0 n_{r0}\alpha_0 + 4k_0 n_{r0} \left(\frac{\pi n_{r1}}{\lambda_0} + i\, \frac{\alpha_1}{2} \right) \cos(2\beta_0 z + \Omega). \tag{4.32}$$

Here, $k_0 = 2\pi/\lambda_0$, where λ_0 is a wavelength in a vacuum.

When the refractive index is constant ($n_{r1} = 0$) and the material is *transparent* ($\alpha_0 = \alpha_1 = 0$), the propagation constant $k(z)$ in (4.32) is given by

$$k(z) = \beta = k_0 n_{r0}. \tag{4.33}$$

In the optical waveguides with corrugations, reflections in Fig. 4.7 (b) couple a forward running wave and a backward one. To express this coupling, we define the *coupling coefficient* κ of the diffraction gratings as

$$\kappa = \frac{\pi n_{r1}}{\lambda_0} + i\, \frac{\alpha_1}{2}, \tag{4.34}$$

which is important to describe the resonance characteristics of the DFB cavity and the DBR.

With the help of (4.33) and (4.34), (4.32) reduces to

$$k(z)^2 = \beta^2 + i\, 2\beta\alpha_0 + 4\beta\kappa \cos(2\beta_0 z + \Omega). \tag{4.35}$$

Substituting (4.35) into a wave equation for the electric field E given by

$$\frac{d^2 E}{dz^2} + k(z)^2 E = 0, \tag{4.36}$$

we obtain

$$\frac{d^2 E}{dz^2} + \left[\beta^2 + i\, 2\beta\alpha_0 + 4\beta\kappa \cos(2\beta_0 z + \Omega) \right] E = 0. \tag{4.37}$$

The electric field $E(z)$, which is a solution of (4.37), is represented by a superposition of a forward running field $E_r(z)$ and a backward one $E_s(z)$ such as

$$\begin{aligned}
E(z) &= E_r(z) + E_s(z), \\
E_r(z) &= R(z) \exp(-i\, \beta_0 z), \\
E_s(z) &= S(z) \exp(i\, \beta_0 z),
\end{aligned} \tag{4.38}$$

where $R(z)$ and $S(z)$ are the field amplitudes of the forward running wave and the backward one, respectively, and both are functions of z. Inserting (4.39) into (4.37) gives the wave equations for R and S as follows:

$$\begin{aligned}
-\frac{dR}{dz} + (\alpha_0 - i\, \delta)R &= i\, \kappa S \exp(-i\Omega), \\
\frac{dS}{dz} + (\alpha_0 - i\, \delta)S &= i\, \kappa R \exp(i\Omega),
\end{aligned} \tag{4.39}$$

where δ is defined as

$$\delta \equiv \frac{\beta^2 - \beta_0{}^2}{2\beta_0} \simeq \beta - \beta_0. \tag{4.40}$$

Here, R and S were assumed to be slowly varying functions of z, and the second derivatives with respect to z were neglected. Because the forward running wave R and the backward one S is coupled by the coupling coefficient κ, (4.39) is called the *coupled wave equation*, and a theory based on (4.39) is referred to as the *coupled wave theory* [16].

As a wavelength for $\delta = 0$, we define *Bragg wavelength* λ_{B} in a vacuum as

$$\lambda_{\mathrm{B}} = \frac{2n_{\mathrm{r}0}\varLambda}{m}, \tag{4.41}$$

where (4.31), (4.33), and (4.40) were used and m is a positive integer designated the *order of diffraction*.

Using constants a and b, which are determined by the boundary conditions, general solutions of (4.37) are given by

$$\begin{aligned}
E_{\mathrm{r}}(z) &= [a\exp(\gamma z) + \rho\exp(-\mathrm{i}\varOmega)\cdot b\exp(-\gamma z)]\exp(-\mathrm{i}\beta_0 z),\\
E_{\mathrm{s}}(z) &= [\rho\exp(\mathrm{i}\varOmega)\cdot a\exp(\gamma z) + b\exp(-\gamma z)]\exp(\mathrm{i}\beta_0 z),
\end{aligned} \tag{4.42}$$

where

$$\gamma^2 = (\alpha_0 - \mathrm{i}\,\delta)^2 + \kappa^2,$$
$$\rho = \frac{-\gamma + (\alpha_0 - \mathrm{i}\,\delta)}{\mathrm{i}\,\kappa}. \tag{4.43}$$

To continue analyzing, it is useful to introduce a *transfer matrix* $\boldsymbol{F}_i$ [17], which is defined as

$$\begin{bmatrix} E_{\mathrm{r}}(0) \\ E_{\mathrm{s}}(0) \end{bmatrix} = \boldsymbol{F}_i \begin{bmatrix} E_{\mathrm{r}}(L_i) \\ E_{\mathrm{s}}(L_i) \end{bmatrix}. \tag{4.44}$$

Here, L_i is the length of a corrugated region and $\boldsymbol{F}_i$ is written as

$$\boldsymbol{F}_i = \begin{bmatrix} F_{11} & F_{12} \\ F_{21} & F_{22} \end{bmatrix},$$

$$F_{11} = \left[\cosh(\gamma L_i) - \frac{\alpha_0 - \mathrm{i}\,\delta}{\gamma}\sinh(\gamma L_i)\right]\exp(\mathrm{i}\,\beta_0 L_i),$$

$$F_{12} = \frac{\mathrm{i}\,\kappa}{\gamma}\sinh(\gamma L_i)\exp[-\mathrm{i}\,(\beta_0 L_i + \varOmega)], \tag{4.45}$$

$$F_{21} = -\frac{\mathrm{i}\,\kappa}{\gamma}\sinh(\gamma L_i)\exp[\mathrm{i}\,(\beta_0 L_i + \varOmega)],$$

$$F_{22} = \left[\cosh(\gamma L_i) + \frac{\alpha_0 - \mathrm{i}\,\delta}{\gamma}\sinh(\gamma L_i)\right]\exp(-\mathrm{i}\,\beta_0 L_i).$$

When multiple regions are connected in series, as shown in Fig. 4.9, the total transfer matrix $\boldsymbol{F}$ is given by a product of transfer matrixes in all regions such as

$$\boldsymbol{F} = \prod_i \boldsymbol{F}_i. \tag{4.46}$$

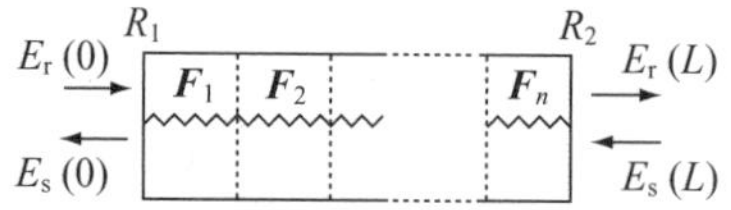

Fig. 4.9. Analytical model for a diffraction grating

If both facets have the power reflectivity of R_1 and R_2, the total transfer matrix $\boldsymbol{F}_{\mathrm{R}}$ is given by

$$\boldsymbol{F}_{\mathrm{R}} = \frac{1}{\sqrt{(1-R_1)(1-R_2)}} \begin{bmatrix} 1 & -\sqrt{R_1} \\ -\sqrt{R_1} & 1 \end{bmatrix} \times \boldsymbol{F} \times \begin{bmatrix} 1 & \sqrt{R_2} \\ \sqrt{R_2} & 1 \end{bmatrix}. \tag{4.47}$$

Let us calculate the power transmissivity T and the power reflectivity R. When we assume that the corrugated region length is L and the input is $E_{\mathrm{r}}(0)$ with $E_{\mathrm{s}}(L) = 0$, the output for the transmission is $E_{\mathrm{r}}(L)$ and that for the reflection is $E_{\mathrm{s}}(0)$. From the definition of the transfer matrix, the power transmissivity T and the power reflectivity R are given by

$$T = \frac{1}{F_{11}{}^* F_{11}}, \quad R = \frac{F_{21}{}^* F_{21}}{F_{11}{}^* F_{11}}. \tag{4.48}$$

Figure 4.10 shows calculated transmission and reflection spectra of a diffraction grating. The horizontal line is $\delta L = \delta \times L$, and the vertical line is the power transmissivity T and the power reflectivity R. Here, it is assumed that the optical waveguide is transparent ($\alpha_0 = \alpha_1 = 0$), $\kappa L = 2$, and $R_1 = R_2 = 0$. In these spectra, there is a low-transmissivity (high-reflectivity) region that is symmetrical about $\delta L = 0$. This region is called the *stop band*, because the transmission is stopped.

4.3.2 Discrete Approach

In the DFB cavity and the DBR, the complex refractive indexes are periodically modulated along the propagation direction of lights. Therefore, they can be analyzed by the *discrete approach*, which has been applied to the *periodic multilayers*.

Figure 4.11 shows a model for analysis where a region with a complex refractive index n_2 and length h_2 and a region with n_3 and h_3 are alternately

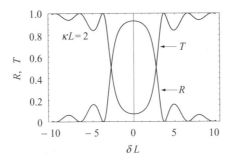

Fig. 4.10. Calculated transmission and reflection spectra of a diffraction grating

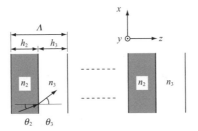

Fig. 4.11. Analytical model in a discrete approach

placed. The angles formed by the interface normal and the light propagation directions are supposed to be θ_2 and θ_3 in the former region and the latter region, respectively.

The relationship between the input and output lights is expressed by a *characteristic matrix* $\boldsymbol{M}_2$, which is defined as

$$\begin{bmatrix} U(0) \\ V(0) \end{bmatrix} = \boldsymbol{M}_2 \begin{bmatrix} U(z) \\ V(z) \end{bmatrix}. \tag{4.49}$$

Here, an electric field $\boldsymbol{E}$ and a magnetic field $\boldsymbol{H}$ are assumed to be expressed by a separation-of-variables procedure. Hence, $U(z)$ and $V(z)$ show dependence of $\boldsymbol{E}$ and $\boldsymbol{H}$ on z. As shown in Fig. 4.11, we choose the light propagation direction as a positive z-axis and the plane of incidence as an xz-plane. As a result, we have $U(z) = E_y(z)$, $V(z) = H_x(z)$ for the TE mode ($E_x = E_z = 0$) and $U(z) = H_y(z)$, $V(z) = -E_x(z)$ for the TM mode ($H_x = H_z = 0$), where a subscript indicates a component along each coordinate.

When we introduce parameters such as

$$\beta_2 = \frac{2\pi}{\lambda_0} n_2 h_2 \cos\theta_2, \quad \beta_3 = \frac{2\pi}{\lambda_0} n_3 h_3 \cos\theta_3,$$

$$p_2 = n_2 \cos\theta_2, \quad p_3 = n_3 \cos\theta_3, \tag{4.50}$$

the characteristic matrix M_2 for the TE mode is written as

$$M_2 = \begin{bmatrix} \cos\beta_2 & -\dfrac{i}{p_2}\sin\beta_2 \\ -i\,p_2\sin\beta_2 & \cos\beta_2 \end{bmatrix} \begin{bmatrix} \cos\beta_3 & -\dfrac{i}{p_3}\sin\beta_3 \\ -i\,p_3\sin\beta_3 & \cos\beta_3 \end{bmatrix}. \tag{4.51}$$

If the number of periods is N, the total characteristic matrix of the optical waveguide M is given by

$$M = M_2{}^N = \begin{bmatrix} m_{11} & m_{12} \\ m_{21} & m_{22} \end{bmatrix}. \tag{4.52}$$

To analyze the DFB cavity and the DBR by the discrete approach, it is useful to put

$$n_2 = n_{r0} + \Delta n, \quad n_3 = n_{r0} - \Delta n,$$
$$h_2 = h_3 = \frac{\Lambda}{2}, \quad \theta_2 = \theta_3 = 0, \tag{4.53}$$

where n_{r0} is an average refractive index of a material; Δn is a shift of refractive index from n_{r0}; Λ is the grating pitch; and $L = N\Lambda$ is the corrugated region length. If $\theta_2 = \theta_3 = 0$, we have common characteristic matrixes for the TE mode and the TM mode. With the help of (4.52), when the outside of the DFB cavity and the DBR is the air, the power transmissivity T and the power reflectivity R are obtained as

$$T = \left| \frac{2}{(m_{11} + m_{12}) + (m_{21} + m_{22})} \right|^2 ,$$
$$R = \left| \frac{(m_{11} + m_{12}) - (m_{21} + m_{22})}{(m_{11} + m_{12}) + (m_{21} + m_{22})} \right|^2 , \tag{4.54}$$

where the boundary conditions at the interfaces for E and H were used. Note that the obtained T and R are derived from *Maxwell's equations*. In this process, we have assumed that E and H are expressed by a separation-of-variables procedure, and approximations have not been used. For more detailed explanations on the characteristic matrix, see Appendix E.

4.3.3 Comparison of Coupled Wave Theory and Discrete Approach

To clarify the application limits of the coupled wave theory and the discrete approach, we will compare the results of these two theories. For brevity, we assume that a material is transparent ($\alpha_0 = \alpha_1 = 0$) and consider the reflectivity at Bragg wavelength ($\delta = 0$) in the first-order diffraction gratings ($m = 1$).

(a) Coupled Wave Theory

From (4.43), the assumption of $\alpha_0 - i\delta = 0$ leads to $\gamma = \pm\kappa$. Substituting this result into (4.45) and (4.48), we obtain the power reflectivity R as

$$R = \tanh^2(\kappa L), \qquad (4.55)$$

where L is the corrugated region length. From (4.55), it is found that κ and L are important to determine the reflectivity of the diffraction gratings.

(b) Discrete Approach

When $\Delta n \ll n_{r0}$ is satisfied, the power reflectivity R is obtained as

$$R = \left[1 - \left(\frac{n_{r0} - \Delta n}{n_{r0} + \Delta n}\right)^{2N}\right]^2 \left[1 + \left(\frac{n_{r0} - \Delta n}{n_{r0} + \Delta n}\right)^{2N}\right]^{-2}, \qquad (4.56)$$

where (4.51)–(4.54) were used. The second terms in both brackets are approximated as

$$\left(\frac{n_{r0} - \Delta n}{n_{r0} + \Delta n}\right)^{2N} \simeq \left(1 - \frac{\Delta n}{n_{r0}}\right)^{4N} = \left[\left(1 - \frac{\Delta n}{n_{r0}}\right)^{n_{r0}/\Delta n}\right]^{4N\Delta n/n_{r0}}$$

$$= \exp\left(-4N\frac{\Delta n}{n_{r0}}\right). \qquad (4.57)$$

With the help of $N\Lambda = L$ and (4.41), the exponent of the right-hand side in (4.57) is written as

$$4N\frac{\Delta n}{n_{r0}} = \frac{8\Delta n}{2n_{r0}\Lambda} L = \frac{8\Delta n}{\lambda_B} L. \qquad (4.58)$$

If we put

$$8\Delta n = 2\pi n_{r1} \qquad (4.59)$$

and substitute (4.59) into (4.58), we have

$$4N\frac{\Delta n}{n_{r0}} = 2\frac{\pi n_{r1}}{\lambda_B} L = 2\kappa L, \qquad (4.60)$$

where (4.34) and the assumption of $\alpha_1 = 0$ were used. Substituting (4.60) into (4.57) results in

$$\left(\frac{n_{r0} - \Delta n}{n_{r0} + \Delta n}\right)^{2N} \simeq e^{-2\kappa L}. \qquad (4.61)$$

Inserting (4.61) into (4.56) leads to

$$R \simeq \left[\frac{1 - e^{-2\kappa L}}{1 + e^{-2\kappa L}} \right]^2 = \tanh^2(\kappa L). \tag{4.62}$$

From (4.55) and (4.62), it is found that the result of the coupled wave theory agrees with that of the discrete approach when $\Delta n \ll n_{r0}$ is satisfied.

It should be noted that the coupled wave theory assumes that complex refractive indexes vary sinusoidally, while the discrete approach presumes that complex refractive indexes change abruptly. Also, in the example for the discrete approach, two layers were alternately placed. Therefore, with an increase in Δn, the results of the coupled wave theory and the discrete approach differ from each other. However, with an increase in the number of layers in one period so that the complex refractive indexes may change sinusoidally in one period, the results of the two theories agree even when Δn is large. Also, the discrete approach can easily analyze any corrugation shape, because one period can be decomposed into many layers. However, because Δn in the DFB cavity and the DBR is on the order of 10^{-3}, the coupled wave theory is widely used to analyze or design the DFB cavity and the DBR.

As shown in Fig. 4.12, a uniform grating in the DFB cavity or the DBR and a periodic multilayer seem different in shape. However, they can be analyzed by a common theory, because the operating principle is common. We should always focus on the essentials, irrespective of superficial differences.

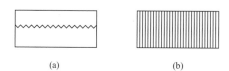

(a)   (b)

Fig. 4.12. (a) Uniform grating and (b) periodic multilayer

4.3.4 Category of Diffraction Gratings

From the viewpoint of the pitch and depth, the diffraction gratings are divided into four groups.

The diffraction grating with uniform pitch and depth, which is shown in Fig. 4.12 (a), is a *uniform grating*. Other diffraction gratings are shown in Fig. 4.13. Figure 4.13 (a) is a *phase-shifted grating* [18,19] whose corrugations shift in its optical waveguide, and this grating is especially important for longitudinal single-mode operations in the DFB-LDs. Figure 4.13 (b) is a *tapered grating* whose corrugation depth is spatially modulated along the propagation direction of the lights. Figure 4.13 (c) is a *chirped grating* whose pitch varies along the propagation direction.

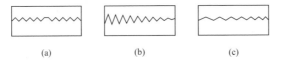

Fig. 4.13. Diffraction gratings used for the DFB cavity and the DBR: (a) phase-shifted, (b) tapered, and (c) chirped

4.3.5 Phase-Shifted Grating

The refractive indexes of the optical fibers change according to the wavelengths of lights, and propagation velocities of lights vary with their wavelengths, which is referred to as *chromatic dispersion*. Hence, if semiconductor lasers show multimode operations, optical pulses broaden in the time domain, and finally adjacent optical pulses overlap each other to limit the transmission of signals. This overlap of adjacent pulses becomes serious with increases in the transmission distance and the signal speed. Therefore, longitudinal single-mode semiconductor lasers are required for long-haul, large-capacity optical fiber communication systems. To achieve stable single-mode operations, DFB-LDs with phase-shifted gratings or gain-coupled gratings have been developed. In the following, we will explain the phase-shifted gratings, which have been used commercially.

(a) Transmission and Reflection Characteristics

Figure 4.14 shows a schematic structure of the phase-shifted grating, in which the corrugation phase is shifted by $-\Delta\Omega$ along the z-axis, and the corrugations in the uniform gratings are illustrated by a broken line. It should be noted that the negative sign of the phase shift is related to the definition of the spatial distribution of the complex refractive index in (4.30).

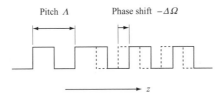

Fig. 4.14. Phase-shifted grating

Let us analyze the phase-shifted gratings using the coupled wave theory. We suppose that both the pitch and the depth are uniform except in the phase-shifted region. Figure 4.15 shows an analytical model, where the diffraction gratings consist of two regions, and the phase shift is introduced as a phase jump at the interface of the two regions.

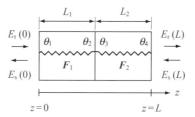

Fig. 4.15. Analytical model for the phase-shifted grating

The transfer matrixes of Regions 1 and 2 are expressed as $\boldsymbol{F}_1$ and $\boldsymbol{F}_2$, respectively. The phase Ω at the left edge of Region 1 is written as θ_1. Then the phase θ_2 at the right edge of Region 1 is given by

$$\theta_2 = \theta_1 + 2\beta_0 L_1. \tag{4.63}$$

Due to the phase shift $\Delta\Omega$, the phase θ_3 at the left edge of Region 2 is obtained as

$$\theta_3 = \theta_2 + \Delta\Omega = \theta_1 + 2\beta_0 L_1 + \Delta\Omega. \tag{4.64}$$

With the help of (4.48), the transmission and reflection characteristics of the phase-shifted gratings are given by the transfer matrix $\boldsymbol{F} = \boldsymbol{F}_1 \times \boldsymbol{F}_2$.

Figure 4.16 shows calculated reflection spectra for $\kappa L = 2$, where the horizontal line is $\delta L = \delta \times L$ and the vertical line is the power reflectivity. Here, it has been assumed that a material is transparent and the facet reflectivity is null. The solid line and broken line correspond to the phase-shifted grating with $-\Delta\Omega = \pi$ and the uniform grating with $-\Delta\Omega = 0$, respectively. The characteristic feature of the phase-shifted grating is that it has a pass band within the stop band, and the phase-shifted DFB-LDs oscillate at this transmission wavelength. The transmission wavelength located within the stop band depends on the phase shift. When the phase shift is π, the transmission wavelength agrees with Bragg wavelength.

(b) Comparison of the Phase-Shifted Grating and the Fabry-Perot Cavity [20]

Comparing the phase-shifted gratings with the Fabry-Perot cavity, we can understand a physical meaning of the relationship between the transmission wavelength and the phase shift.

Figure 4.17 schematically shows the Fabry-Perot cavity where the power reflectivity of the mirror is R_0 and the cavity length is L. From (4.12), the resonance condition of the Fabry-Perot cavity is given by

$$L = n \frac{\lambda_0}{2n_{\mathrm{r}}}. \tag{4.65}$$

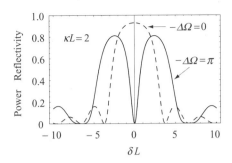

Fig. 4.16. Reflection spectrum

Here, n is a positive integer, n_r is the effective refractive index of a material, and λ_0 is a light wavelength in a vacuum. Figure 4.18 shows the power reflectivity (reflectance) R_0 of the *mirror* and the power transmissivity (transmittance) T of the *Fabry-Perot cavity* as a function of a wavelength λ. It should be noted that R_0 is independent of λ if the dispersions are neglected, and there are sharp peaks for all the wavelengths satisfying the resonance condition (4.65).

Fig. 4.17. Fabry-Perot cavity

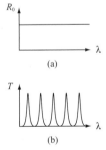

Fig. 4.18. Resonance characteristics of the Fabry-Perot cavity: (a) reflectance of a mirror and (b) transmittance of a cavity

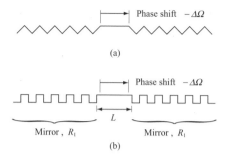

Fig. 4.19. Phase-shifted grating: (a) saw-toothed grating and (b) rectangular grating

Figure 4.19 schematically shows the phase-shifted gratings with a saw-toothed shape and a rectangular shape. Grating shapes affect the value of κ, but the concepts on the grating pitch are common to both structures. Therefore, to focus on only the grating pitch, we will consider the rectangular shape in the following.

The phase shift $-\Delta\Omega$ is defined as illustrated in Fig. 4.19. Such a phase-shifted grating is regarded as the Fabry-Perot cavity with length L and the two mirrors, which have wavelength-dependent reflectivities. In this case, a relation between L and $\Delta\Omega$ is expressed as

$$L = \frac{\Lambda}{2} + \frac{|\Delta\Omega|}{2\beta_0} = \frac{\Lambda}{2}\left(1 + \frac{|\Delta\Omega|}{\pi}\right), \tag{4.66}$$

where (4.31) was used. From (4.66), it is found that the cavity length L changes with the phase shift $-\Delta\Omega$, which varies a resonance (transmission) wavelength according to (4.12). For example, when the phase shift is π, (4.41) and (4.66) give

$$L = \Lambda = m\frac{\lambda_B}{2n_{r0}}, \tag{4.67}$$

where m is a positive integer called the order of diffraction and n_{r0} is the averaged refractive index. From (4.67), it is found that the resonance wavelength for $-\Delta\Omega = \pi$ is a product of Bragg wavelength and a positive integer. For the first-order grating ($m = 1$), the resonance wavelength is Bragg wavelength.

Figure 4.20 shows the power reflectivity R_1 of the *mirror* in Fig. 4.19 and the power transmissivity T of the *phase-shifted grating (cavity)* with $-\Delta\Omega = \pi$ as a function of a wavelength. The power reflectivity R_1 of the mirror depends on a wavelength λ, and only the selected wavelength region has high reflectivity. Therefore, only a resonance wavelength located in the high reflectivity region is selectively multireflected in the cavity. Also, because the cavity length L is on the order of a wavelength, the mode spacing (free

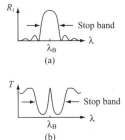

(a)

(b)

Fig. 4.20. Resonance characteristics of the phase-shifted grating: (a) reflectance of a mirror and (b) transmittance of a cavity

spectral range) is so large that only one transmission peak exists within the stop band.

As described earlier, the phase-shifted grating can be qualitatively explained as the Fabry-Perot cavity. The only difference between the phase-shifted grating and the Fabry-Perot cavity is the dependence of the mirror reflectivities on a wavelength.

Note: Definition of the Phase Shift [21]

There are two definitions of the phase shift in the phase-shifted grating, as shown in Fig. 4.21. In Fig. 4.21 (a), Region 1 ($z < 0$) and Region 2 ($z > 0$) shift symmetrically with respect to $z = 0$, while in Fig. 4.21 (b), Region 1 does not move and Region 2 ($z > 0$) shifts at $z = 0$.

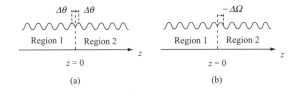

(a)   (b)

Fig. 4.21. Definition of the phase shift

Using the definition in Fig. 4.21 (a), the refractive index $n_r(z)$ in Region 1 is expressed as

$$n_r(z) = n_{r0} + n_{r1} \cos(2\beta_0 z + \Delta\theta),$$ (4.68)

whereas the refractive index in Region 2 is written as

$$n_r(z) = n_{r0} + n_{r1} \cos(2\beta_0 z - \Delta\theta).$$ (4.69)

When we use the definition in Fig. 4.21 (b), the refractive index in Region 2 is written as

$$n_r(z) = n_{r0} + n_{r1} \cos(2\beta_0 z + \Delta\theta + \Delta\Omega), \qquad (4.70)$$

where (4.68) was used. Because (4.69) and (4.70) represent the same refractive index, the two phase shifts $\Delta\Omega$ and $\Delta\theta$ are related as

$$\Delta\theta + \Delta\Omega = 2m\pi - \Delta\theta, \qquad (4.71)$$

which reduces to

$$2\Delta\theta + \Delta\Omega = 2m\pi, \qquad (4.72)$$

where m is an integer.

These two definitions of the phase shift are used in various journal papers or books. Therefore, we should be careful in discussing a value of the phase-shift.

4.3.6 Fabrication of Diffraction Gratings

In DFB-LDs and DBR-LDs, oscillation wavelengths are just or in the vicinity of Bragg wavelengths. The oscillation wavelength for the light sources of the optical fiber transmission systems is 1.3 μm in which the *dispersion* of the optical fiber is the smallest, or 1.55 μm where the *absorption loss* of the optical fiber is the lowest. Because the effective refractive index n_{r0} of the semiconductor lasers is nearly 3.2, pitches of the first-order grating Λ are approximately 0.2 μm for a wavelength of 1.3 μm and 0.24 μm for 1.55 μm from (4.41). Corrugation depth is about 0.1 μm = 100 nm just after the grating fabrication and is reduced to several tens of nanometers after epitaxial growth of semiconductor layers on the diffraction grating. This reduction in the grating depth is caused by thermal decomposition of the grating surface during heating prior to the epitaxial growth.

To fabricate such fine diffraction gratings with high accuracy, *holographic exposure* [22–30], *electron-beam exposure* [31], and *X-ray exposure* [32] systems have been developed.

(a) Holographic Exposure

Holographic exposure systems use interference of two coherent laser beams to make *interference fringe patterns*. The obtained interference fringe patterns are transferred to the *photoresists* coated on the substrates via developing the photoresists. Finally, the substrates are etched with the patterned photoresists as the *etching masks*.

Figure 4.22 shows a principle of the holographic exposure, where solid lines show wave fronts of two lightwaves and closed circles indicate the points in which the light intensity is enhanced due to interference. As the wave fronts propagate toward the substrate, the closed circles propagate along the broken lines. Consequently, the photoresist regions indicated by arrows are exposed. The grating pitch Λ is determined by the spacing of the broken lines.

Figure 4.23 shows the angles of incidence θ_1 and θ_2, which are formed by the normals of the substrate plane and the propagation directions of the two laser beams. The grating pitch Λ is given by

$$\Lambda = \frac{\lambda_e}{\sin\theta_1 + \sin\theta_2}, \tag{4.73}$$

where λ_e is a wavelength of the incident laser beams.

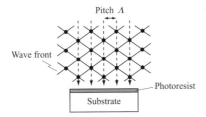

Fig. 4.22. Principle of holographic exposure

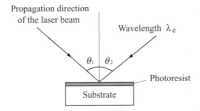

Fig. 4.23. Angle of incidence in holographic exposure

Figure 4.24 shows an example of the holographic exposure systems. As the light sources, He-Cd lasers ($\lambda_e = 441.6\,\text{nm}$, $325\,\text{nm}$) or Ar ion lasers ($\lambda_e = 488\,\text{nm}$) are widely used. A single laser beam emitted from the light source is divided into two laser beams by a beam splitter, and these two laser beams are expanded. These expanded beams are collimated and are finally incident on the photoresists coated on the substrates. To achieve fine patterns with high accuracy, we need high uniformity in the wave fronts; low fluctuations in the optical paths; high stability in the wavelength, the phase, and the light intensity of the laser beams; and precise control of the angle of incidence. For these purposes, the holographic exposure systems must be isolated from mechanical vibrations and air flow.

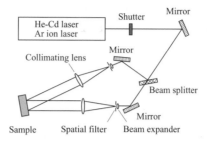

Fig. 4.24. Holographic exposure system

(b) Electron Beam Exposure

The electron beam exposure systems scan electron beams on the photoresist in a vacuum, as shown in Fig. 4.25. Scanning of the electron beams and transfer of the samples are controlled by computers, which results in flexible patterns. The problem is that exposure time is long. For example, it takes about 10 hours to draw patterns of the diffraction grating on a 1-cm square substrate.

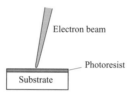

Fig. 4.25. Electron beam exposure

(c) X-Ray Exposure

Wavelengths of X-rays are short, which leads to small diffraction angles. Hence, X-rays are suitable to transfer fine patterns of a photomask to the photoresist, as shown in Fig. 4.26. To obtain sufficient X-ray intensity for exposure, *synchrotron radiation* is often used, which needs a huge plant. Also, highly reliable photomasks for the X-ray exposure systems have not yet been developed.

(d) Fabrication of Phase-Shifted Grating

The electron beam and X-ray exposure systems are suitable to fabricate various patterns. These systems, however, have problems on costs and productivity. As a result, the holographic exposure systems, which have high productivity with low cost, have attracted a lot of interest. As will be explained

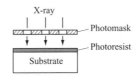

Fig. 4.26. X-ray exposure

in Chapter 6, the most stable single-mode operations are obtained when the phase-shift is $|\Delta\Omega| = \pi$, which corresponds to a shift in the length $\Lambda/2$ in the first-order grating. In other words, the top and bottom of the corrugations are reversed in the optical waveguide.

From the viewpoint of the *reverse* of the corrugations, *positive* and *negative* photoresists are simultaneously exposed. In a positive photoresist, exposed areas are removed by development, whereas in a negative photoresist, unexposed areas are removed. Therefore, by selectively forming the positive photoresist and the negative photoresist on the substrate, we can obtain a pattern and its reverse on the same plane. Figure 4.27 shows an example of this method, where SiN is used to prevent chemical reactions between the positive photoresist and the negative one [23].

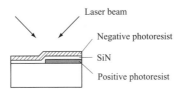

Fig. 4.27. Holographic exposure using positive and negative photoresists

From the viewpoint of the *shift in the pitch*, wave fronts of the laser beams are shifted, as shown in Figs 4.28 and 4.29.

In Fig. 4.28, a material with a larger refractive index than air is selectively placed on the surface of the photoresist to shift both wave fronts of the two incident laser beams. The lights incident on this material are refracted by Snell's law, which leads to a change in the exposed positions. In Fig. 4.28 (a) [25], the phase-shift plate is put on the photoresist. If there is a tiny air gap such as the order of $1\,\mu m$ between the phase-shift plate and the photoresist, exposed patterns are heavily degraded due to multireflections of the laser beams. Hence, precise position control of the phase-shift plate and the photoresist is required. In Fig. 4.28 (b) [26], in contrast, the phase-shift layer is coated on a buffer layer, which automatically results in no air gap between the phase-shift layer and the buffer layer.

In Fig. 4.29, an optical element such as a phase-shift plate or a hologram is inserted in the optical path for one laser beam to shift one of the wave fronts of the two incident laser beams. In Fig. 4.29 (a) [27], the wave fronts are disturbed due to diffractions at steps in the phase-shift plate, which reduces the grating formed area. Figure 4.29 (b) [28] uses a hologram to generate a required phase shift on the photoresist, and there are no distortions in the wave fronts. However, a highly reliable hologram has not yet been developed.

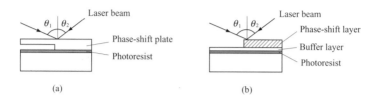

Fig. 4.28. Holographic exposure shifting both wavefronts of the two incident laser beams

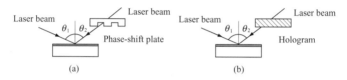

Fig. 4.29. Holographic exposure shifting one of the wavefronts of the two incident laser beams

Figure 4.30 shows a replica method [29] where one laser beam is incident on a replica of a master phase-shifted grating. The fringe patterns, which are formed by interference of the transmitted light and the diffracted light from the replica, are transferred to a photoresist.

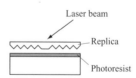

Fig. 4.30. Replica method

Because of reproducibility, productivity, a large tolerance in lithography conditions, and costs, the phase-shift method in Fig. 4.28 (b) was first applied to manufacture phase-shifted gratings. Later, the replica method in Fig. 4.30 was also commercially used. Recently, electron beam exposure systems have been used in some factories.

5 Fundamentals of Semiconductor Lasers

5.1 Key Elements in Semiconductor Lasers

As shown in Fig. 5.1, a laser is composed of an active material, which has the optical gain, and an optical resonator, which feeds back lights by its reflectors. In *semiconductor lasers*, the *active layers* generate the spontaneous emission and amplify a fraction of the spontaneous emission by the stimulated emission. As the optical resonators, the Fabry-Perot cavities, the ring cavities, the DFB cavities, and the DBRs are used. In this chapter, we study fundamental characteristics of semiconductor lasers using the Fabry-Perot LDs, which are shown in Fig. 5.2. Dynamic single-mode lasers such as the DFB-LDs will be discussed in Chapter 6.

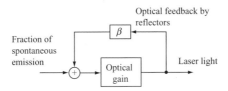

Fig. 5.1. Key elements in the lasers

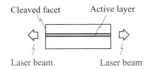

Fig. 5.2. Fabry-Perot LD

In the following, we will explain the *optical resonator*, the *pn-junction*, and the *double heterostructure* (DH), which are indispensable elements in semiconductor lasers.

5.1.1 Fabry-Perot Cavity

As explained in Chapter 4, the Fabry-Perot cavity comprises two parallel mirrors. In semiconductor lasers, cleaved facets such as $\{011\}$ or $\{0\bar{1}1\}$ surfaces are used as mirrors, as shown in Fig. 5.2. When a light is normally incident on a facet, the power reflectivity R_0 is given by

$$R_0 = \left(\frac{n_{rt} - 1}{n_{rt} + 1} \right)^2 , \tag{5.1}$$

where n_{rt} and 1 are the refractive indexes of the semiconductors and the air, respectively.

When n_{rt} is 3.5, R_0 is about 31%. It should be noted that the cleaved facets are flat with the order of atomic layers, and the surface is much smoother than the wavelengths of lights. Therefore, the cleaved facets function as mirrors with high accuracy. In order to control the reflectivities or to protect the facets, dielectric films are often coated on the cleaved facets.

5.1.2 pn-Junction

In semiconductor lasers, in order to inject the carriers into the active layers, the active layers are placed inside the pn-junctions. Therefore, the active layer is sandwiched by the *p-cladding layer* and the *n-cladding layer*, as shown in Fig. 5.3. Applying a *forward bias* voltage, which is positive on the p-side and negative on the n-side, across this pn-junction, the electrons are injected from the n-cladding layer to the active layer, and the holes are injected from the p-cladding layer to the active layer, as shown in Fig. 5.4. As explained in Chapter 2, when the population inversion is generated by the carrier injection, net stimulated emission is obtained.

Fig. 5.3. Cross section of a pn-junction in semiconductor lasers

Note that the impurities are often *undoped* in the active layers to achieve high radiation efficiency. However, in the active layers, there are background carriers whose concentrations depend on epitaxial growth methods. Therefore, the active layers are not ideal *intrinsic* semiconductors.

If the impurities are doped in the active layers, the injected carriers combine with the impurities. Therefore, the *carrier lifetime* is reduced, and the modulation speed is enhanced (see Sects. 5.11 and 5.12). However, the recombinations of the injected carriers and the impurities do not contribute

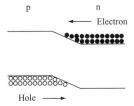

Fig. 5.4. Carrier injection in a pn-junction under a forward bias

to laser transitions, which decreases the radiation efficiency. Thus, the active layers are sometimes intentionally doped to achieve high-speed modulations so long as the radiation efficiency is not highly degraded.

5.1.3 Double Heterostructure

The *heterojunction* is a junction consisting of different materials in which materials with different compositions are also categorized. In contrast, the junction composed of common material is called the *homojunction*.

The bandgap energies in the semiconductors depend on the constituent elements and the compositions. As a result, the heterostructures have the *energy barriers* at the junction interfaces, and these energy barriers confine the carriers to the well layers. To achieve efficient recombinations of the electrons and the holes, these carriers have to be confined to the active layers. Therefore, the heterostructures are formed at both interfaces of the active layer. Such a structure is designated the *double heterostructure* because there are double heterojunctions.

Figure 5.5 shows the distributions of the energy and the refractive index of the double heterostructure. At the junction interfaces, there are the *band offsets* ΔE_c for the conduction band and ΔE_v for the valence band, as shown in Fig. 5.5 (a). Under a forward bias, the holes are injected from the p-cladding layer to the active layer, and the electrons are injected from the n-cladding layer to the active layer. The energy barrier for the holes is ΔE_v at the interface of the n-cladding layer and the active layer; that for the electrons is ΔE_c at the interface of the p-cladding layer and the active layer. In many semiconductors, their refractive indexes increase with a decrease in the bandgap energies. Hence, the refractive index of the active layer n_a is usually greater than that of the p-cladding layer n_p and that of the n-cladding layer n_n. As a result, a light is efficiently confined to the active layer, which results in a high light amplification rate.

As described earlier, the double heterostructure confines both the carriers and the light to the active layer. Therefore, the double heterostructure is indispensable to achieve excellent characteristics in semiconductor lasers. It should be noted that the first continuous wave (CW) laser oscillation at room temperature was achieved by the double heterostructure [33,34].

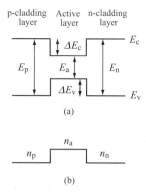

Fig. 5.5. Double heterostructure: (a) energy of electrons and (b) distribution of refractive index

5.2 Threshold Gain

We explain the *threshold gain*, which is the optical gain required for laser oscillation. From (4.9), the transmitted light intensity I_t and the incident light intensity I_0 are related as

$$\frac{I_t}{I_0} = \frac{T_1 T_2 G_{s0}}{(1 - G_{s0}\sqrt{R_1 R_2})^2 + 4G_{s0}\sqrt{R_1 R_2}\sin^2(n_r k_0 L)}, \tag{5.2}$$
$$G_{s0} = \exp(2g_E L) = \exp(gL),$$

where g_E is the amplitude gain coefficient for the electric field and g is the optical power gain coefficient.

Oscillation is a state in which there is an output without an input from outside. As a result, the oscillation condition is given by $I_0 = 0$ and $I_t > 0$ in (5.2). Therefore, at the oscillation condition, the denominator in (5.2) is 0, and (5.2) goes to infinity. Hence, the oscillation condition for the Fabry-Perot LDs is expressed as

$$\begin{array}{ll} \text{Resonance condition}: & \sin(n_r k_0 L) = 0, \\ \text{Gain condition}: & 1 - G_{s0}\sqrt{R_1 R_2} = 0. \end{array} \tag{5.3}$$

5.2.1 Resonance Condition

From (5.3), the *resonance condition* is written as

$$\frac{n_r \omega L}{c} = n_r k_0 L = n\pi, \tag{5.4}$$

where n is a positive integer. Using a wavelength in a vacuum λ_0, (5.4) reduces to

$$L = n\frac{\lambda_0}{2n_{\mathrm{r}}}, \tag{5.5}$$

which is the same as the resonance condition of the Fabry-Perot cavity explained in Chapter 4. Note that laser oscillation starts at the resonance wavelength nearest to the gain peak.

5.2.2 Gain Condition

From (5.3), the *gain condition* is obtained as

$$1 - G_{\mathrm{s0}}\sqrt{R_1 R_2} = 1 - \sqrt{R_1 R_2}\exp(gL) = 0. \tag{5.6}$$

As a result, the optical power gain coefficient g is written as

$$g = \frac{1}{L}\ln\frac{1}{\sqrt{R_1 R_2}}, \tag{5.7}$$

where the right-hand side is called the *mirror loss*.

As described in Chapter 3, the guided modes propagate in the optical waveguides while confined to the guiding layer (active layer), and the fields of the guided modes penetrate into the p-cladding layer and the n-cladding layer, as shown in Fig. 5.6. Hence, the light sees the optical losses in the p-cladding layer and the n-cladding layer. Therefore, we have to consider the optical power gain coefficient g for the entire region where the lights exist.

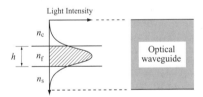

Fig. 5.6. Distribution of light intensity in the optical waveguide

Using the optical confinement factors, the optical power gain coefficient, and the optical power loss coefficients of the layers in the optical waveguide, we can approximately write the optical power gain coefficient g of the optical waveguide as

$$g = \Gamma_{\mathrm{a}}g_{\mathrm{a}} - \Gamma_{\mathrm{a}}\alpha_{\mathrm{a}} - \Gamma_{\mathrm{p}}\alpha_{\mathrm{p}} - \Gamma_{\mathrm{n}}\alpha_{\mathrm{n}}. \tag{5.8}$$

Here, Γ_{a}, Γ_{p}, and Γ_{n} are the optical confinement factors of the active layer, the p-cladding layer, and the n-cladding layer, respectively; g_{a} is the optical power gain coefficient of the active layer; and α_{a}, α_{p}, and α_{n} are the optical power loss coefficients of the active layer, the p-cladding layer, and the n-cladding layer, respectively. It should be noted that the exact g value is obtained by solving eigenvalue equations, which include complex refractive indexes.

Introducing the *internal loss* as

$$\alpha_i = \Gamma_a \alpha_a + \Gamma_p \alpha_p + \Gamma_n \alpha_n \tag{5.9}$$

and substituting (5.9) into (5.8), we have

$$g = \Gamma_a g_a - \alpha_i, \tag{5.10}$$

where $\Gamma_a g_a$ is the modal gain. Inserting (5.10) into (5.7) results in

$$\Gamma_a g_a = \alpha_i + \frac{1}{L} \ln \frac{1}{\sqrt{R_1 R_2}} = \alpha_i + \frac{1}{2L} \ln \frac{1}{R_1 R_2}, \tag{5.11}$$

which is the threshold gain of the Fabry-Perot LDs. In this equation, the left-hand side is the threshold gain, and the right-hand side is the total loss. Therefore, (5.11) indicates that laser oscillation takes place when the optical gain is equal to the total loss, which is the sum of the internal loss and the mirror loss. In InGaAsP/InP LDs for lightwave communications, $\Gamma_a g_a$ is 50–60 cm^{-1} because of $R_1 = R_2 = 32\%$, $L = 250$–300 μm, and $\alpha_i = 10$–20 cm^{-1}. When the electric currents are injected into semiconductor lasers, laser oscillation starts at the threshold current I_{th}, which satisfies (5.11).

5.3 Radiation Efficiency

Figure 5.7 schematically shows the current versus light output (*I-L*) characteristics of the semiconductor lasers. As shown in Fig. 5.7, when the injection current I exceeds the threshold current I_{th}, laser beams are emitted outward. To evaluate radiation efficiency of the semiconductor lasers, the *slope efficiency* and the *external differential quantum efficiency* are used.

5.3.1 Slope Efficiency

The *slope efficiency* S_{dj} (in units of mW/mA or W/A) per facet is defined as the ratio of the increase in light intensity ΔP_j ($j = 1, 2$) to the increase in injection current ΔI, which is given by

$$S_{dj} = \frac{\Delta P_j}{\Delta I}. \tag{5.12}$$

The slope efficiency for the total light output is obtained as

$$S_{d,tot} = \frac{\Delta P_1 + \Delta P_2}{\Delta I} = \frac{\Delta P}{\Delta I}, \tag{5.13}$$

where ΔP is an increase in the total light intensity.

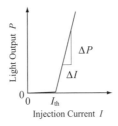

Fig. 5.7. Current versus light output characteristics

5.3.2 External Differential Quantum Efficiency

The *external differential quantum efficiency* η_d (in no units) is defined as the number of photons emitted outward per injected carrier. The external differential quantum efficiency η_d for the total light output is given by

$$\eta_d = \frac{\Delta P}{\hbar \omega} \div \left(\frac{\Delta I}{e}\right) = \frac{\Delta P}{\Delta I} \frac{e}{\hbar \omega} = S_{d,tot} \frac{e}{\hbar \omega} , \qquad (5.14)$$

where ω is an angular frequency of the light, $\hbar$ is Dirac's constant, and e is the elementary charge.

As shown in (5.11), the total loss is a sum of the internal loss and the mirror loss. For a reference system placed outside the optical cavity, the mirror loss indicates the light emission rate from the optical cavity. As a result, using the *internal quantum efficiency* η_i, which is defined as the number of photons emitted inside the optical cavity per injected carrier, the external differential quantum efficiency η_d is expressed as

$$\eta_d = \eta_i \frac{(\text{Mirror Loss})}{(\text{Total Loss})} = \eta_i \frac{\frac{1}{2L} \ln \frac{1}{R_1 R_2}}{\alpha_i + \frac{1}{2L} \ln \frac{1}{R_1 R_2}} = \eta_i \frac{\ln \frac{1}{R_1 R_2}}{2\alpha_i L + \ln \frac{1}{R_1 R_2}}, \qquad (5.15)$$

where the optical losses at the facets due to the absorption or the scattering were assumed to be negligible. For $\eta_i = 100\%$, $L = 300\,\mu\text{m}$, $R_1 = R_2 = 32\%$, and $\alpha_i = 20\,\text{cm}^{-1}$, we have $\eta_d = 66\%$.

5.3.3 Light Output Ratio from Facets

We consider the light outputs P_1 and P_2 from the two facets. As shown in Fig. 5.8, we assume that the light intensities in the vicinity of the facets inside the optical cavity are P_a, P_b, P_c, and P_d; the power reflectivities of the facets are R_1 and R_2; and the power transmissivities of the facets are T_1 and T_2. Here, the arrows indicate the propagation directions of the lights. Among the light intensities in a steady state, we have a relationship such as

$$P_a = R_1 P_d, \quad P_b = e^{gL} P_a, \quad P_c = R_2 P_b, \quad P_d = e^{gL} P_c,$$
$$P_1 = T_1 P_d, \quad P_2 = T_2 P_b, \tag{5.16}$$

where g is the optical power gain coefficient and L is the cavity length. Deleting P_k (k = a, b, c, d) from (5.16), we have

$$\frac{P_1}{P_2} = \frac{T_1 \sqrt{R_2}}{T_2 \sqrt{R_1}}. \tag{5.17}$$

Substituting $P = P_1 + P_2$ into (5.16) leads to

$$P_1 = \frac{T_1 \sqrt{R_2}}{T_1 \sqrt{R_2} + T_2 \sqrt{R_1}} P, \quad P_2 = \frac{T_2 \sqrt{R_1}}{T_1 \sqrt{R_2} + T_2 \sqrt{R_1}} P. \tag{5.18}$$

Hence, the external differential quantum efficiencies for each light output η_{d1} and η_{d2} are written as

$$\eta_{d1} = \frac{T_1 \sqrt{R_2}}{T_1 \sqrt{R_2} + T_2 \sqrt{R_1}} \eta_d, \quad \eta_{d2} = \frac{T_2 \sqrt{R_1}}{T_1 \sqrt{R_2} + T_2 \sqrt{R_1}} \eta_d, \tag{5.19}$$

where η_d is the external differential quantum efficiency for the total light output. When the optical losses at the facets are negligibly small, we have $T_1 = 1 - R_1$ and $T_2 = 1 - R_2$.

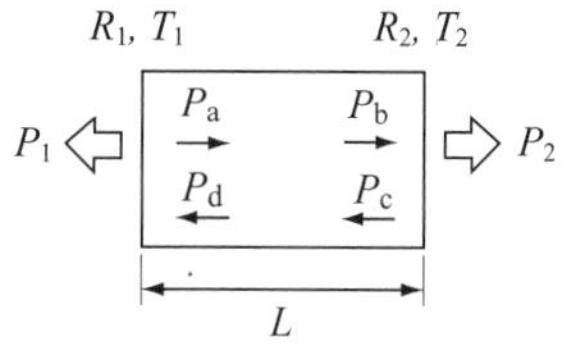

Fig. 5.8. Light intensities inside and outside the Fabry-Perot cavity

5.4 Current versus Light Output (*I-L*) Characteristics

We study the current versus light output (*I-L*) characteristics above and below the threshold. With an increase in the injection current into semiconductor lasers, the carrier concentration in the active layer is enhanced. When the carrier concentration exceeds the threshold carrier concentration, laser oscillation starts and the light output drastically increases compared to below the threshold. This change in light output is considered to be a change in the photon density. Therefore, we analyze *I-L* characteristics by introducing the *rate equations* on the *carrier concentration* and the *photon density* in the active layer. Dependence of *I-L* on temperature is also briefly described.

5.4.1 Rate Equations

If we assume that the electron concentration n is equal to the hole concentration p, we can write the *rate equations* for the carrier concentration n and the photon density S of the laser light as

$$\frac{dn}{dt} = \frac{J}{ed} - G(n)S - \frac{n}{\tau_n}, \qquad (5.20)$$

$$\frac{dS}{dt} = G(n)S - \frac{S}{\tau_{ph}} + \beta_{sp}\frac{n}{\tau_r}. \qquad (5.21)$$

Here, J is the injection current density, which is an electric current flowing through a unit area; e is the elementary charge; d is the active layer thickness; $G(n)$ is the amplification rate due to the stimulated emission; τ_n is the *carrier lifetime*; τ_{ph} is the *photon lifetime*; β_{sp} is the *spontaneous emission coupling factor*; and τ_r is the *radiative recombination lifetime* due to the spontaneous emission.

In (5.20), $J/(ed)$ is an increased rate of the carrier concentration in the active layer; $-G(n)S$ shows a consumption rate of the carrier concentration due to the stimulated emission and is proportional to the photon density S; and $-n/\tau_n$ expresses a decay rate of the carrier concentration in the carrier lifetime τ_n.

In (5.21), $G(n)S$ shows an increased rate of the photon density S due to the stimulated emission; $-S/\tau_{ph}$ is a decreased rate of the photon density inside the optical cavity due to the absorption and light emission toward the outside of the optical cavity; and $\beta_{sp}\, n/\tau_r$ represents a coupling rate of spontaneously emitted photons to the lasing mode, which is a resonance mode of the cavity.

Here, we explain $G(n)$, τ_n, τ_{ph}, and β_{sp} in more detail. With the transparent carrier concentration n_0, in which a material is transparent, we can approximately write $G(n)$ as

$$G(n) = \Gamma_a g_0(n - n_0), \qquad (5.22)$$

where g_0 is the differential gain coefficient and Γ_a is the optical confinement factor of the active layer.

Using the *radiative recombination lifetime* τ_r and the *nonradiative recombination lifetime* τ_{nr}, we can express the carrier lifetime τ_n as

$$\frac{1}{\tau_n} = \frac{1}{\tau_r} + \frac{1}{\tau_{nr}}. \qquad (5.23)$$

The radiative recombination lifetime τ_r is determined by the spontaneous emission and is not affected by the stimulated emission. The nonradiative recombination lifetime τ_{nr} is related to the recombinations of the injected carriers and the defects or the impurities, which do not emit lights.

The *photon lifetime* τ_{ph} is the time during which the photons stay in the optical cavity; it is expressed as

$$\frac{1}{\tau_{\mathrm{ph}}} = \frac{c}{n_{\mathrm{r}}}\left(\alpha_{\mathrm{i}} + \frac{1}{2L}\ln\frac{1}{R_1 R_2}\right), \tag{5.24}$$

where n_{r} is the effective refractive index and c is the speed of light in a vacuum.

Here, let us derive (5.24). If we put

$$\alpha_{\mathrm{t}} = \alpha_{\mathrm{i}} + \frac{1}{2L}\ln\frac{1}{R_1 R_2}, \tag{5.25}$$

we can write a differential equation for the photon density S with respect to a position z as

$$\frac{\mathrm{d}S}{\mathrm{d}z} = -\alpha_{\mathrm{t}}\, S, \tag{5.26}$$

where the light is assumed to propagate toward a positive z-axis. As a result, a derivative of the photon density S with respect to a time t is given by

$$\frac{\mathrm{d}S}{\mathrm{d}t} = \frac{\mathrm{d}z}{\mathrm{d}t}\frac{\mathrm{d}S}{\mathrm{d}z} = \frac{c}{n_{\mathrm{r}}}\frac{\mathrm{d}S}{\mathrm{d}z} = -\frac{c}{n_{\mathrm{r}}}\alpha_{\mathrm{t}}\, S \equiv -\frac{S}{\tau_{\mathrm{ph}}}. \tag{5.27}$$

From (5.25) and (5.27), the photon lifetime τ_{ph} in (5.24) is obtained.

The *spontaneous emission coupling factor* β_{sp} is defined as

$$\beta_{\mathrm{sp}} = \frac{(\text{spontaneous emission coupling rate to the lasing mode})}{(\text{total spontaneous emission rate})}. \tag{5.28}$$

When the spontaneous emission spectrum is assumed to be *Lorentzian* with the center angular frequency ω_0 and the FWHM $\Delta\omega$, the spontaneous emission coupling rate to the lasing mode per unit time and unit volume r_{sp} is given by

$$r_{\mathrm{sp}} = r_{\mathrm{sp0}}\frac{(\Delta\omega/2)^2}{(\omega - \omega_0)^2 + (\Delta\omega/2)^2}, \tag{5.29}$$

where r_{sp0} is a coefficient.

To calculate the total spontaneous emission rate, we consider the number of modes $\mathrm{d}N$ with two polarizations, which exist in a volume V, a solid angle for propagation direction $\mathrm{d}\Omega$; and an angular frequency range $\mathrm{d}\omega$. When the distribution of the modes is continuous, as in a free space, $\mathrm{d}N$ is given by

$$\mathrm{d}N = V\, m(\omega)\, \mathrm{d}\omega\,\frac{\mathrm{d}\Omega}{4\pi} = V\,\frac{n_{\mathrm{r}}^3\omega^2}{\pi^2 c^3}\,\mathrm{d}\omega\,\frac{\mathrm{d}\Omega}{4\pi}, \tag{5.30}$$

where (2.26) was used. From (5.29) and (5.30), the total spontaneous emission rate R_{sp} is obtained as

$$R_{\mathrm{sp}} = \int r_{\mathrm{sp}}\, dN = r_{\mathrm{sp}0}\, \frac{V}{2\pi}\left(\frac{n_r}{c}\right)^3 {\omega_0}^2\, \Delta\omega. \tag{5.31}$$

Using (5.28), (5.29), and (5.31), we can express the spontaneous emission coupling factor β_{sp} for the angular frequency ω_0 (a wavelength in a vacuum λ_0) as

$$\beta_{\mathrm{sp}} = \Gamma_a \frac{r_{\mathrm{sp}}}{R_{\mathrm{sp}}} = \Gamma_a \frac{2\pi}{V}\left(\frac{c}{n_r}\right)^3 \frac{1}{{\omega_0}^2 \Delta\omega} = \frac{\Gamma_a}{4\pi^2 {n_r}^3 V}\frac{{\lambda_0}^4}{\Delta\lambda}, \tag{5.32}$$

where Γ_a is the optical confinement factor of the active layer and $\Delta\lambda$ is the FWHM in units of wavelength. From (5.32), it is found that the spontaneous emission coupling factor β_{sp} increases with a decrease in the mode volume V and the spectral linewidth $\Delta\lambda$.

5.4.2 Threshold Current Density

Let us calculate the *threshold current density* J_{th} using the rate equations.

First, we consider the rate equations below the threshold, where net stimulated emission is negligible and $S = 0$. Therefore, (5.20) reduces to

$$\frac{dn}{dt} = \frac{J}{ed} - \frac{n}{\tau_n}. \tag{5.33}$$

In a steady state ($d/dt = 0$), from (5.33), the carrier concentration n is given by

$$n = \frac{J}{ed}\tau_n. \tag{5.34}$$

When the carrier concentration n increases from 0 to the *threshold carrier concentration* n_{th}, we expect that (5.34) is still satisfied at the threshold. As a result, the threshold current density J_{th} is expressed as

$$J_{\mathrm{th}} = \frac{ed}{\tau_n} n_{\mathrm{th}}. \tag{5.35}$$

From (5.35), it is found that a small n_{th} and a long τ_n lead to a low J_{th}. Because the optical confinement factor Γ_a of the active layer depends on the active layer thickness d, the threshold current density n_{th} is a function of d and there exists an optimum d value to achieve the lowest J_{th}.

Secondly, we calculate the threshold carrier concentration n_{th} using the rate equations above the threshold. In usual semiconductor lasers, the spontaneous emission coupling factor β_{sp} is on the order of 10^{-5}. Therefore, as the first approximation, we neglect the term $\beta_{\mathrm{sp}} n/\tau_r$.

Because (5.21) is valid for any S value in a steady state, we have

$$G(n) = \Gamma_a g_0 (n - n_0) = \frac{1}{\tau_{\mathrm{ph}}}, \tag{5.36}$$

where (5.22) was used. Substituting (5.11) and (5.24) into (5.36), we also obtain

$$G(n) = \frac{c}{n_{\mathrm{r}}} \Gamma_{\mathrm{a}} g_{\mathrm{a}}. \tag{5.37}$$

From (5.36), the carrier concentration n in a steady state is given by

$$n = \frac{1}{\Gamma_{\mathrm{a}} g_0 \tau_{\mathrm{ph}}} + n_0. \tag{5.38}$$

Because (5.38) is satisfied even at the threshold, the threshold carrier concentration n_{th} is written as

$$n_{\mathrm{th}} = \frac{1}{\Gamma_{\mathrm{a}} g_0 \tau_{\mathrm{ph}}} + n_0. \tag{5.39}$$

In semiconductor lasers, changes in the cavity length, the facet reflectivities, and the refractive indexes during laser operation are small, and the right-hand sides of (5.38) and (5.39) are considered to be constant. Therefore, above the threshold, the carrier concentration n is clamped on the threshold carrier concentration n_{th}. Hence, $G(n)$ is constant above the threshold, as far as the gain saturation and the coupling of the spontaneous emission to the lasing mode are neglected.

Substituting (5.39) into (5.35), we have

$$J_{\mathrm{th}} = \frac{ed}{\tau_n} n_{\mathrm{th}} = \frac{ed}{\tau_n} \left(\frac{1}{\Gamma_{\mathrm{a}} g_0 \tau_{\mathrm{ph}}} + n_0 \right),$$

$$A = \frac{ed}{\tau_n} n_0, \quad B = \frac{ed}{\tau_n} \frac{1}{\Gamma_{\mathrm{a}} g_0 \tau_{\mathrm{ph}}}, \tag{5.40}$$

where it is clearly shown that the threshold current density J_{th} depends on the optical confinement factor Γ_{a}. Figure 5.9 shows calculated results of Γ_{a} for $\mathrm{Al}_x\mathrm{Ga}_{1-x}\mathrm{As/GaAs}$ double heterostructures. As shown in Fig. 5.9, with an increase in the active layer thickness d, Γ_{a} is enhanced. Note that Γ_{a} is proportional to d^2 when d is small.

Figure 5.10 shows the threshold current density J_{th} as a function of the active layer thickness d. It is found that J_{th} takes a minimum value when d is approximately $0.1\,\mu\mathrm{m}$. In Fig. 5.10, A is a current density, which is required to obtain the population inversion and is proportional to the active layer thickness d as in (5.40). On the other hand, B is a current density, in which the optical gain balances the loss in the optical cavity. For a thin active layer, B is inversely proportional to d, because Γ_{a} is proportional to d^2. Because J_{th} is given by $A + B$, there exists an optimum d value to obtain a minimum J_{th}.

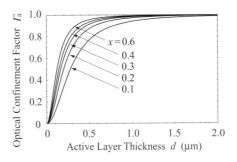

Fig. 5.9. Optical confinement factor

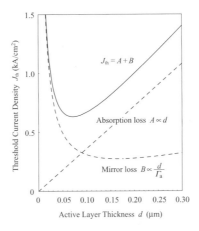

Fig. 5.10. Dependence of the threshold current density J_{th} on the active layer thickness d

5.4.3 Current versus Light Output (I-L) Characteristics in CW Operation

(a) Without Coupling of Spontaneous Emission to the Lasing Mode

Let us examine changes in the carrier concentration n and the photon density S with the injection current density J. When the spontaneous emission coupling factor β_{sp} is small, coupling of the spontaneous emission to the lasing mode can be neglected.

Below the threshold, with an increase in J, the carrier concentration n increases according to (5.34), but the photon density S is 0. Above the threshold, n does not increase any more and remains at the threshold carrier concentration n_{th}, while S drastically increases with J, because the excess carriers $(n - n_{th})$ are converted to photons.

From (5.20), the steady-state photon density S above the threshold is obtained as

$$S = \frac{1}{G(n)}\left(\frac{J}{ed} - \frac{n_{\text{th}}}{\tau_n}\right). \tag{5.41}$$

Substituting (5.35) and (5.36) into (5.41) results in

$$S = \frac{\tau_{\text{ph}}}{ed}(J - J_{\text{th}}). \tag{5.42}$$

In summary, dependence of the carrier concentration n and the photon density S on the injection current density J is expressed as

in $J < J_{\text{th}}$:

$$n = \frac{J}{ed}\tau_n, \quad S = 0; \tag{5.43}$$

in $J \geq J_{\text{th}}$:

$$n = \frac{J_{\text{th}}}{ed}\tau_n, \quad S = \frac{\tau_{\text{ph}}}{ed}(J - J_{\text{th}}). \tag{5.44}$$

Figure 5.11 illustrates the calculated results of (5.43) and (5.44). It is clearly shown that the carrier concentration n is clamped on n_{th} above the threshold current density J_{th}.

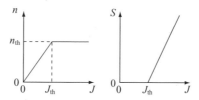

Fig. 5.11. Carrier concentration n and photon density S when coupling of the spontaneous emission to the lasing mode is neglected

(b) With Coupling of Spontaneous Emission to the Lasing Mode

For brevity, we assume that the nonradiative recombination is negligible, which leads to $\tau_r \approx \tau_n$. In a steady state, when the coupling of the spontaneous emission to the lasing mode is included, (5.20) and (5.21) reduce to

$$\frac{J}{ed} = \Gamma_a g_0 (n - n_0) S + \frac{n}{\tau_n}, \tag{5.45}$$

$$\frac{S}{\tau_{\text{ph}}} = \Gamma_a g_0 (n - n_0) S + \beta_{\text{sp}} \frac{n}{\tau_n}, \tag{5.46}$$

where (5.22) was used. Therefore, the carrier concentration n and the photon density S are given by

$$n = \frac{n_{\text{th}}}{2(1 - \beta_{\text{sp}})} \left(X - \sqrt{X^2 - Y} \right), \tag{5.47}$$

$$S = \frac{\beta_{\text{sp}}}{\Gamma_{\text{a}} g_0 \tau_n} \frac{X - \sqrt{X^2 - Y}}{2(1 - \beta_{\text{sp}}) - \left(X - \sqrt{X^2 - Y} \right)}, \tag{5.48}$$

where

$$X = 1 + \frac{J}{J_{\text{th}}} - \beta_{\text{sp}} \frac{n_0}{n_{\text{th}}}, \tag{5.49}$$

$$Y = 4(1 - \beta_{\text{sp}}) \frac{J}{J_{\text{th}}}. \tag{5.50}$$

Figure 5.12 shows the calculated results of (5.47) and (5.48), where solid and broken lines correspond to $\beta_{\text{sp}} > 0$ and $\beta_{\text{sp}} = 0$, respectively. Coupling of the spontaneous emission to the lasing mode lowers n and enhances S, which results in a vague threshold. When $\beta_{\text{sp}} < 1$, however, the emitted lights below the threshold are incoherent spontaneous emissions or amplified spontaneous emissions.

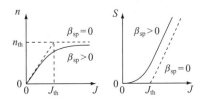

Fig. 5.12. Carrier concentration n and photon density S when coupling of the spontaneous emission to the lasing mode is included

5.4.4 Dependence of *I-L* on Temperature

Generally, with a rise in a temperature, the threshold current density J_{th} increases and the external differential quantum efficiency η_{d} decreases, as shown in Fig. 5.13. Here, the horizontal line and the vertical line show the injection current and light output, respectively.

Dependence of the threshold current density J_{th} on a temperature is empirically expressed as

$$J_{\text{th}} = J_{\text{th}0} \exp \left(\frac{T_{\text{j}}}{T_0} \right), \tag{5.51}$$

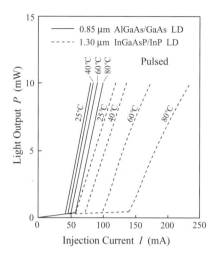

Fig. 5.13. Dependence of *I-L* on temperature

where J_{th0} is a coefficient, T_j is the temperature in the active layer or the junction temperature, and T_0 is the *characteristic temperature*, which indicates dependence of the threshold current density on the temperature.

A large characteristic temperature T_0 seems to result in a small dJ_{th}/dT_j, which indicates a good semiconductor laser. However, we must be careful about evaluating the temperature characteristics of semiconductor lasers by T_0, because a larger J_{th} leads to a greater T_0 when dJ_{th}/dT_j is constant. Therefore, only when semiconductor lasers with a common J_{th} at the same temperature are compared, can T_0 be an appropriate index.

Figure 5.14 shows examples of the characteristic temperature T_0 for an AlGaAs/GaAs LD and an InGaAsP/InP LD in pulsed operations. In an AlGaAs/GaAs LD with an oscillation wavelength of 0.85 μm, T_0 is approximately 160 K from 25°C to 80°C. It is considered that an increase in J_{th} with an increment in T_j is caused by broadening the gain spectrum and the *overflow* of the carriers over the heterobarriers. To reduce the overflow of the carriers over the heterobarriers, we must increase the band offset ΔE_g between the active layer and the cladding layers. In general, the band offset ΔE_g should be larger than 0.3 eV to suppress a drastic increase in J_{th} with a rise in T_j.

The external quantum efficiency η_d decreases with an increase in J_{th}, because the threshold carrier concentration n_{th} increases with J_{th}, which enhances the *free carrier absorption* (see Appendix F).

In an InGaAsP/InP LD with an oscillation wavelength of 1.3 μm, T_0 is approximately 70 K from 25°C to 65°C. This T_0 is lower than that of an AlGaAs/GaAs LD because of efficient overflow of the carriers due to the

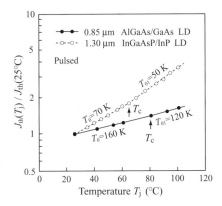

Fig. 5.14. Characteristic temperature T_0

light effective mass of the electrons and the nonradiative recombinations due to the *Auger processes* and the *valence band absorptions*.

The effective masses of the electrons are $0.070m$ for AlGaAs with a bandgap wavelength of $0.85\,\mu m$, and $0.059m$ for InGaAsP with a bandgap wavelength of $1.3\,\mu m$, where m is the electron mass in a vacuum.

The Auger processes are schematically shown in Fig. 5.15 where C, H, L, and S indicate the conduction band, the heavy hole band, the light hole band, and the split-off band, respectively. In the Auger processes, there are two processes, CHSH and CHCC. In the CHSH process, the emitted energy due to a recombination of electron 1 in the conduction band (C) and hole 2 in the heavy hole band (H) excites electron 3 in the split-off band (S) to the heavy hole band (H). In the CHCC process, the energy emitted due to a recombination of electron 1 in the conduction band (C) and hole 2 in the heavy hole band (H) pumps electron 3 in the conduction band (C) to a higher energy state 4 in the conduction band. These processes are three-body collision processes, and a recombination rate R_A for the Auger processes is given by

$$R_A = C_p n p^2 + C_n n^2 p, \tag{5.52}$$

where C_p and C_n are the Auger coefficients for the CHSH and CHCC processes, respectively.

In the valence band absorptions, an electron in the split-off band absorbs a light generated by a recombination of an electron in the conduction band and a hole in the heavy hole band. These absorption processes are shown in Fig. 5.16, where an electron is excited to the heavy hole band and the acceptor level.

With an increase in temperature, the internal quantum efficiency η_i decreases due to the Auger processes and the external quantum efficiency η_d is lowered due to the valence band absorptions and the free carrier absorption.

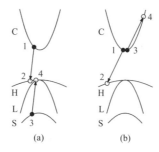

Fig. 5.15. Auger processes: (a) CHSH and (b) CHCC

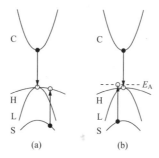

Fig. 5.16. Valence band absorption: electron excited to (a) heavy hole band and (b) the acceptor level

Therefore, light emission efficiency is reduced with an increase in temperature.

5.5 Current versus Voltage (*I-V*) Characteristics

In a steady state, (5.20) and (5.21) reduce to

$$S = -\beta_s \frac{n}{\tau_r} \frac{1}{G(n) - \tau_{\text{ph}}^{-1}}, \tag{5.53}$$

$$I = eV_A \left[G(n)S + \frac{n}{\tau_n} \right]. \tag{5.54}$$

Here, I is the injection current and $V_A = S_A d$ is the volume of the active layer in which S_A is the area of the active layer and d is the active layer thickness.

A flowing current in semiconductor lasers consists of the diffusion current, the drift current, and the recombination current. Here, it is assumed that the radiative recombination is dominant, and the diffusion and drift currents are neglected, as shown in (5.54).

The carrier concentration n is approximately given by

$$n = n_{\mathrm{i}} \exp\left(\frac{eV}{2k_{\mathrm{B}}T}\right), \tag{5.55}$$

where V is a voltage across the pn-junction and n_{i} is the intrinsic carrier concentration, which is expressed as

$$n_{\mathrm{i}} = 2\left(\frac{2\pi k_{\mathrm{B}}T}{h^2}\right)^{3/2} (m_{\mathrm{e}}m_{\mathrm{h}})^{3/4} \exp\left(-\frac{E_{\mathrm{g}}}{2k_{\mathrm{B}}T}\right). \tag{5.56}$$

Here, k_{B} is Boltzmann's constant, T is an absolute temperature, h is Planck's constant, m_{e} and m_{h} are the effective masses of the electron and the hole, respectively, and E_{g} is the bandgap energy of the active layer.

From (5.53)–(5.55), a relation between the injection current I and the voltage V is given by

$$I = eV_{\mathrm{A}}\left[-\beta_{\mathrm{s}}\frac{n_{\mathrm{i}}e^{eV/2k_{\mathrm{B}}T}}{\tau_{\mathrm{r}}}\frac{G(n_{\mathrm{i}}e^{eV/2k_{\mathrm{B}}T})}{G(n_{\mathrm{i}}e^{eV/2k_{\mathrm{B}}T}) - \tau_{\mathrm{ph}}^{-1}} + \frac{n_{\mathrm{i}}e^{eV/2k_{\mathrm{B}}T}}{\tau_n}\right], \tag{5.57}$$

which is illustrated in Fig. 5.17. In this figure, solid, dotted, and broken lines correspond to the photon lifetimes $\tau_{\mathrm{ph}} = 1\,\mathrm{ps}$, $2\,\mathrm{ps}$, and $3\,\mathrm{ps}$, respectively. Here, a sum of a contact resistance and a bulk resistance is assumed to be $4\,\Omega$, which is typical in conventional semiconductor lasers. Other physical parameters that are used in this calculation are $\beta_{\mathrm{s}} = 10^{-5}$, $n_{\mathrm{i}} = 2.7 \times 10^{11}\,\mathrm{cm}^{-3}$, $\tau_{\mathrm{r}} = \tau_n = 1\,\mathrm{ns}$, $T = 293.15\,\mathrm{K}$ (20°C), $(\partial G/\partial n)_{n=n_{\mathrm{th}}} = 2.5 \times 10^{-6}\,\mathrm{cm}^3/\mathrm{s}$, and $V_{\mathrm{A}} = 40\,\mu\mathrm{m}^3 = 4 \times 10^{-11}\,\mathrm{cm}^3$. It is also supposed that the transparent carrier concentration is $n_0 = 0.6\,n_{\mathrm{th}}$, the effective refractive index is 3.5, and the group velocity v_{g} is $8.57 \times 10^9\,\mathrm{cm/s}$.

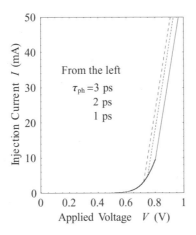

Fig. 5.17. Current versus voltage ($I - V$) characteristics

5.6 Derivative Characteristics

We can use *derivative measurements* to precisely detect the threshold current. In the following, relationships among the threshold current, the *derivative light output*, and the *derivative electrical resistance* will be explained.

5.6.1 Derivative Light Output

Using the photon density S in the active layer, we can write the internal light intensity P as

$$P = h\nu v_{\mathrm{g}} S_{\mathrm{B}} S, \tag{5.58}$$

where h is Planck's constant, ν is an oscillation frequency, v_{g} is a group velocity of the photons, and S_{B} is a beam area. From (5.58), the derivative light output with respect to the injection current $\mathrm{d}P/\mathrm{d}I$ is written as

$$\frac{\mathrm{d}P}{\mathrm{d}I} = h\nu v_{\mathrm{g}} S_{\mathrm{B}} \frac{\mathrm{d}S}{\mathrm{d}I} = h\nu v_{\mathrm{g}} S_{\mathrm{B}} \frac{\partial S}{\partial n}\frac{\partial n}{\partial I}. \tag{5.59}$$

Figure 5.18 shows calculated light output and derivative light output as a function of the injection current where (5.53), (5.54), (5.58), and (5.59) were used. The used parameters are the same as in Fig. 5.17. A sharp rise in $\mathrm{d}P/\mathrm{d}I$ clearly indicates the threshold current.

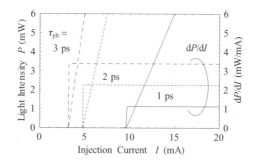

Fig. 5.18. Current versus light output and derivative light output

5.6.2 Derivative Electrical Resistance

The derivative resistance $\mathrm{d}V/\mathrm{d}I$ is expressed as

$$\frac{\mathrm{d}V}{\mathrm{d}I} = \frac{\partial V}{\partial n}\frac{\partial n}{\partial I}. \tag{5.60}$$

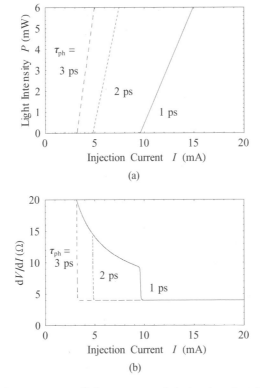

Fig. 5.19. Current versus light output and derivative electrical resistance

Figure 5.19 shows calculated I-L and I-dV/dI characteristics where (5.55)–(5.57) and (5.60) were used. The parameters used are the same as in Figs. 5.17 and 5.18.

As shown in Figs. 5.19 (a) and (b), the I-dV/dI curves have kinks at the threshold currents. Therefore, the threshold currents are determined by the derivative electrical resistance dV/dI, and this method will be especially useful for semiconductor ring or disk lasers with extremely low light output.

5.7 Polarization of Light

The Fabry-Perot LDs, which have *bulk* active layers, oscillate in the *TE mode*. As explained in Chapter 1, the bulk active layers do not have particular quantum mechanical axes. As a result, the optical gains for the bulk active layers have values averaged with respect to all angles and are independent of the polarization of lights. However, the facet reflectivities depend on the polarization of lights.

Figure 5.20 (a) schematically shows reflection at a facet of a semiconductor laser in a zigzag model. We suppose that the effective refractive index of the optical waveguide is n'_A and the refractive index of the outside of the optical waveguide is n_B. This reflection is also considered as the reflection at the interface of a material with the refractive index $n_A = n'_A / \cos \theta$ and that with the refractive index n_B, as shown in Fig. 5.20 (b).

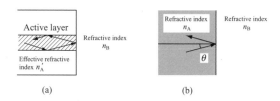

(a)   (b)

Fig. 5.20. Reflection at a facet

When the angle of incidence is θ, *Fresnel formulas* give the power reflectivities R_{TE} and R_{TM} as

$$R_{TE} = \left| \frac{n_A \cos \theta - \sqrt{n_B{}^2 - n_A{}^2 \sin^2 \theta}}{n_A \cos \theta + \sqrt{n_B{}^2 - n_A{}^2 \sin^2 \theta}} \right|^2 , \tag{5.61}$$

$$R_{TM} = \left| \frac{n_B{}^2 \cos \theta - n_A \sqrt{n_B{}^2 - n_A{}^2 \sin^2 \theta}}{n_B{}^2 \cos \theta + n_A \sqrt{n_B{}^2 - n_A{}^2 \sin^2 \theta}} \right|^2 , \tag{5.62}$$

where subscripts indicate the polarization of lights. Figure 5.21 shows the power reflectivity R when laser beams are emitted from GaAs with the refractive index $n_A = 3.6$ to the air with $n_B = 1$. As found in Fig. 5.21, a relation $R_{TE} \geq R_{TM}$ is kept for all values of θ. Hence, from (5.11), the threshold gains for the TE modes are lower than those of the TM modes, which results in a start of lasing in the TE modes. With an increase in θ, i.e., with an increase in the order of modes, R_{TE} is enhanced. Therefore, the threshold gains of higher-order modes are smaller than those of lower-order modes. However, in order to minimize the threshold current density, the active layer thickness d should be approximately 0.1 μm, as shown in Fig. 5.10. In such a small d, the higher-order modes are cut off, and only the fundamental mode oscillates.

Note that the light intensity ratio of the TE and TM modes is approximately 1 : 1 below the threshold, because the light emitted below the threshold is the spontaneous emission.

5.8 Parameters and Specifications

There are trade-offs among characteristics of semiconductor lasers, which are determined by the internal loss α_i, the power reflectivity at a facet R,

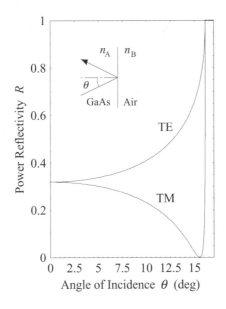

Fig. 5.21. Reflectivity for TE and TM modes

the cavity length L, the active layer thickness d, and so on. For example, a long L and a large R result in a low threshold gain $\Gamma_a g_a$, as shown in (5.11). However, a long L leads to a large volume of the active layer V_A, which results in a large threshold current $I_{th} = J_{th} \times V_A$. Also, a large R leads to a low power transmissivity T, which results in a low light output as found in (5.18). Therefore, we need to design semiconductor lasers to satisfy specifications according to applications.

5.9 Two-Mode Operation

Up to now, for brevity, we have considered single-mode operations and have neglected the gain saturation. Here, we treat two-mode operations by including the self-saturation and cross-saturation of the optical gains [35–40].

The rate equations for the photon densities S_1 and S_2 in two-mode operations are given by

$$\dot{S}_1 = (\alpha_1 - \beta_1 S_1 - \theta_{12} S_2) S_1, \qquad (5.63)$$

$$\dot{S}_2 = (\alpha_2 - \beta_2 S_2 - \theta_{21} S_1) S_2, \qquad (5.64)$$

where $\alpha_i \equiv G_i - 1/\tau_{\mathrm{ph}i}$ is the net amplification rate, G_i is the amplification rate, $\tau_{\mathrm{ph}i}$ is the photon lifetime for the ith mode, β_i is the self-saturation coefficient, and θ_{ij} is the cross-saturation coefficient $(i, j = 1, 2)$. In a steady state $(\mathrm{d}/\mathrm{d}t = 0)$, we obtain

$$\beta_1 S_1 + \theta_{12} S_2 = \alpha_1 : L_1, \tag{5.65}$$
$$\beta_2 S_2 + \theta_{21} S_1 = \alpha_2 : L_2, \tag{5.66}$$

where L_1 and L_2 indicate lines in Fig. 5.22. From small variations analysis, the coupling constant C is defined as

$$C \equiv \frac{\theta_{12}\theta_{21}}{\beta_1\beta_2}. \tag{5.67}$$

According to the value of C, the coupling strengths are classified into three regions as

$$C < 1 \quad \text{weak coupling,}$$
$$C = 1 \quad \text{neutral coupling,}$$
$$C > 1 \quad \text{strong coupling.}$$

Relationships between S_1 and S_2 are summarized in Fig. 5.22. As found in (5.63) and (5.64), we have $\dot{S}_i < 0$ in a region above line L_i ($i = 1, 2$), and $\dot{S}_i > 0$ in a region below line L_i ($i = 1, 2$). According to a sign of $\dot{S}_i$, a stable point of the photon density S_i is determined.

Figures 5.22 (a) and (b) show the weak coupling ($C < 1$); in (a), for $\beta_1\alpha_2/\theta_{21} < \alpha_1$ only mode 1 oscillates and oscillation of mode 2 is inhibited, and in (b), for $\theta_{12}\alpha_2/\beta_2 < \alpha_1 < \beta_1\alpha_2/\theta_{21}$ mode 1 and mode 2 simultaneously oscillate, which is only allowed in the weak coupling. If $\alpha_1 < \theta_{12}\alpha_2/\beta_2$, only mode 2 oscillates to prevent mode 1 from oscillating.

Figures 5.22 (c) and (d) show the neutral coupling ($C = 1$), and lines L_1 and L_2 are parallel. For $\beta_2\alpha_1/\theta_{12} = \theta_{21}\alpha_1/\beta_1 > \alpha_2$, as shown in Figs. 5.22 (c) and (d), only mode 1 oscillates and mode 2 cannot oscillate. If $\beta_2\alpha_1/\theta_{12} = \theta_{21}\alpha_1/\beta_1 < \alpha_2$, only mode 2 oscillates oscillation of mode 1 suppressed.

Figures 5.22 (e) and (f) show the strong coupling ($C > 1$); in (e), for $\alpha_2 < \beta_2\alpha_1/\theta_{12}$ only mode 1 oscillates, and in (f), for $\beta_2\alpha_1/\theta_{12} < \alpha_2 < \theta_{21}\alpha_1/\beta_1$, mode 1 or mode 2 oscillates according to the initial values of S_1 and S_2. For example, a change in the refractive index of the optical waveguide by an applied voltage or an injection current leads to a bistable operation between S_1 and S_2, which is only observed in the strong coupling. If $\theta_{21}\alpha_1/\beta_1 < \alpha_2$, only mode 2 oscillates.

5.10 Transverse Modes

The *transverse modes*, or *lateral modes*, show the light intensity distributions along the axes perpendicular to the cavity axis, which determine the shapes of the laser beams. The transverse modes are highly dependent on the structures of the optical waveguides, and the *vertical transverse modes* display the light intensity distributions along the axes perpendicular to the

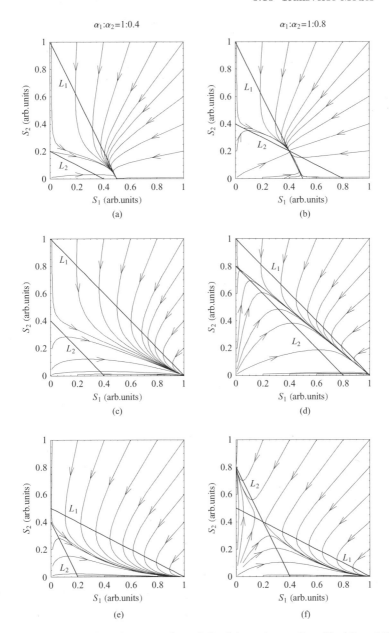

Fig. 5.22. Relationships between S_1 and S_2: (a) weak coupling $(\theta_{ij}/\beta_i = 0.5)$, (b) neutral coupling $(\theta_{ij}/\beta_i = 1)$, and (c) strong coupling $(\theta_{ij}/\beta_i = 2)$

active layer plane, whereas the *horizontal transverse modes* exhibit the light intensity distributions along the axes parallel to the active layer plane.

To evaluate the transverse modes, we usually use *near-field patterns* and *far-field patterns*, which are shown in Fig. 5.23.

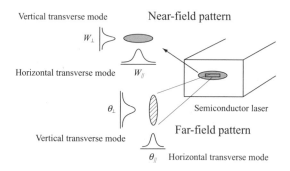

Fig. 5.23. Near-field and far-field patterns

The near-field pattern is the light intensity distributions on a facet; its indexes are the length of the emission region $W_\parallel$ and $W_\perp$ illustrated in Fig. 5.23. Usually, the active layer thickness d is approximately $0.1\,\mu m$ or less to minimize the threshold current density, while the active layer width is on the order of $2\,\mu m$ or more to achieve high reproducibility with high accuracy. As a result, the near-field pattern is asymmetric, which is long along the axis parallel to the active layer plane and short along the axis perpendicular to the active layer plane.

The far-field pattern is the light intensity distributions at a position that is far enough from the facet. As shown in Fig. 5.23, its indexes are the radiation angles $\theta_\parallel$ and $\theta_\perp$, which are independent of the distance between the facet and the detector. This far-field pattern is considered to be a *diffracted pattern* of the near-field pattern if the near-field pattern is regarded as a *slit*. With a decrease in the size of the slit, the size of the diffracted patterns increases. Therefore, the far-field patterns are large for small near-field patterns and small for large near-field patterns. Because of asymmetry in the near-field patterns, the far-field patterns are also asymmetric with small horizontal transverse modes and large vertical ones.

When we couple a laser beam to an optical component such as a lens or an optical fiber, we would like to achieve a large coupling efficiency. For this purpose, a symmetric laser beam with a narrow radiation angle is required. In this respect, conventional semiconductor lasers are not optimized, and the optical coupling loss is minimized by optimizing the optical components or the optical coupling systems. Recently, however, semiconductor lasers with the optical waveguides whose thickness or width is graded along the cavity axes have been demonstrated to enhance the optical coupling efficiencies. Vertical cavity surface emitting lasers are also suitable to achieve high optical coupling efficiency because their emitted beams have symmetric circular shapes with narrow radiation angles (see Chapter 6).

5.10.1 Vertical Transverse Modes

(a) Guided Modes

The vertical transverse modes present the light intensity distributions along the axes perpendicular to the active layer plane. Because the double heterostructures are adopted in semiconductor lasers, index guiding is established along the axis perpendicular to the active layer plane. The guiding condition is represented by the eigenvalue equation for the transverse resonance condition.

Here, we briefly review the key points of the eigenvalue equation, which was explained in Chapter 3. Figure 5.24 shows an optical waveguide, in which the guiding layer with the refractive index n_f and the thickness h is sandwiched by the cladding layer with the refractive index n_c and the substrate with the refractive index n_s where $n_f > n_s \geq n_c$.

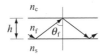

Fig. 5.24. Cross section of an optical waveguide

Because the Fabry-Perot LDs with bulk active layers oscillate in the TE modes, we consider the eigenvalue equation for the TE modes. Using a wave number in a vacuum k_0, the normalized frequency or the normalized waveguide thickness V is defined as

$$V = k_0 h \sqrt{n_f{}^2 - n_s{}^2}. \tag{5.68}$$

With the help of the effective refractive index N, the normalized waveguide refractive index b_{TE} is defined as

$$b_{TE} = \frac{N^2 - n_s{}^2}{n_f{}^2 - n_s{}^2}. \tag{5.69}$$

Also, we introduce the asymmetry measure a_{TE} as

$$a_{TE} = \frac{n_s{}^2 - n_c^2}{n_f{}^2 - n_s{}^2}. \tag{5.70}$$

Using these normalized parameters, we can express the normalized eigenvalue equation for the TE modes as

$$V\sqrt{1 - b_{\mathrm{TE}}} = m\pi + \tan^{-1}\sqrt{\frac{b_{\mathrm{TE}}}{1 - b_{\mathrm{TE}}}} + \tan^{-1}\sqrt{\frac{a_{\mathrm{TE}} + b_{\mathrm{TE}}}{1 - b_{\mathrm{TE}}}}, \qquad (5.71)$$

where m is a nonnegative integer, which is called the order of modes.

The important concept in designing the optical waveguides is the cutoff condition in which the guided modes do not exist. When the angle of incidence (= angle of reflection) in the guiding layer θ_{f} is equal to the critical angle θ_{fs}, propagated lights are not confined to the guiding layer and a fraction of the lights are emitted to the substrate. In this case, we have $N = n_{\mathrm{s}}$, which results in $b_{\mathrm{TE}} = 0$ from (5.69). As a result, from (5.71), the normalized frequency V_m for the mth-order mode to be cut off is obtained as

$$V_m = m\pi + \tan^{-1}\sqrt{a_{\mathrm{TE}}}. \qquad (5.72)$$

For $V_m < V < V_{m+1}$, the guided modes from the zeroth- to the mth-order modes exist. In a symmetric optical waveguide with $n_{\mathrm{s}} = n_{\mathrm{c}}$ where $a_{\mathrm{TE}} = 0$ in (5.70), (5.72) reduces to

$$V_m = m\pi. \qquad (5.73)$$

From (5.68) and (5.73), the cutoff guiding layer thickness h_{c} in a symmetric optical waveguide is expressed as

$$h_{\mathrm{c}} = \frac{m\pi}{k_0\sqrt{n_{\mathrm{f}}^2 - n_{\mathrm{s}}^2}} = \frac{m\lambda_0}{2\sqrt{n_{\mathrm{f}}^2 - n_{\mathrm{s}}^2}}, \qquad k_0 = \frac{2\pi}{\lambda_0}, \qquad (5.74)$$

where λ_0 is a wavelength of a light in a vacuum. From (5.74), with increases in h_{c} and $n_{\mathrm{f}}^2 - n_{\mathrm{s}}^2$, higher-order modes with large ms can be guided in the optical waveguides.

Using V and b, the optical confinement factor Γ_{f} in the symmetric optical waveguide is given by

$$\Gamma_{\mathrm{f}} = \frac{V\sqrt{b} + 2b}{V\sqrt{b} + 2}. \qquad (5.75)$$

(b) Laser Oscillation in Higher-Order Modes

(i) Active Layer Thickness $d \gtrsim h_{\mathrm{c}}$

From (5.74) and (5.75), a relationship of the optical confinement factor Γ_m for the mth-order mode and Γ_{m-1} for the $(m-1)$th-order mode is written as

$$\Gamma_m < \Gamma_{m-1}. \qquad (5.76)$$

When the active layer thickness d is slightly larger than h_{c}, a difference in the reflectivities between the adjacent higher-order modes is small. Therefore, from (5.40), the threshold current density $J_{th,m}$ for the mth-order mode and $J_{th,m-1}$ for the $(m-1)$th-order mode are related as

$$J_{th,m} > J_{th,m-1}, \tag{5.77}$$

which leads to a laser oscillation in the $(m-1)$th-order mode, not the mth-order mode.

(ii) Active Layer Thickness $d \gg h_c$

As shown in Fig. 5.21, higher-order TE modes have larger reflectivities than lower-order ones. Therefore, when the active layer thickness d is much larger than h_c, a laser oscillation in the mth-order mode takes place. For an AlGaAs/GaAs LD with $n_f = 3.6$, $(n_f - n_s)/n_f = 5\%$, and $\lambda_0 = 0.85\,\mu$m, the cutoff guiding layer thickness h_{cm} for the mth-order mode is obtained as

$$h_{c1} = 0.38\,\mu\text{m}, \quad h_{c2} = 0.76\,\mu\text{m}, \quad h_{c3} = 1.13\,\mu\text{m}.$$

In this case, according to the active layer thickness d, the following laser oscillations in higher-order modes take place:

$$d > 0.66\,\mu\text{m} : \text{ first-order mode}$$
$$d > 0.98\,\mu\text{m} : \text{ second-order mode}$$
$$d > 1.30\,\mu\text{m} : \text{ third-order mode}$$

In conventional semiconductor lasers, to minimize the threshold current density, the active layer thickness d is approximately $0.1\,\mu$m, as shown in Fig. 5.10. Therefore, the vertical transverse modes are *fundamental modes* $(m = 0)$.

(c) Near-Field Pattern and Far-Field Pattern

Let us consider the vertical size of the near-field patterns $W_\perp$ and of the far-field patterns $\theta_\perp$. When the active layer thickness d is larger than a light wavelength in a material, $W_\perp$ shrinks with a decrease in d, because the light emission region narrows. However, when d is less than a light wavelength in a material, $W_\perp$ increases with a decrease in d, because the guided lights highly penetrate into the cladding layer and the substrate.

As described before, the far-field pattern is considered to be a diffracted pattern of the near-field pattern. As a result, the far-field pattern is large for a small near-field pattern and is small for a large near-field pattern.

Figure 5.25 shows the calculated results of $W_\perp$ and $\theta_\perp$ for the fundamental TE mode $(m = 0)$. The horizontal line is the active layer thickness d, and Δ in the inset is $(n_f - n_s)/n_f$. For $d = 0.1\,\mu$m, the radiation angle $\theta_\perp$ is approximately $30°$, which is not good enough for optical coupling. However, this value is much smaller than those of LDs below the threshold or LEDs. This is because the mode for the laser light is the guided mode, while the spontaneously emitted lights include not only the guided modes but also the radiation modes, which have various polarizations and propagation directions. In LEDs, to achieve a relatively narrow radiation angle, monolithic or hybrid lenses are formed on their light emission surfaces.

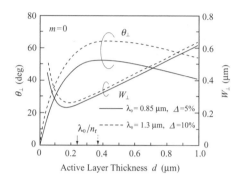

Fig. 5.25. Sizes of near-field and far-field patterns

5.10.2 Horizontal Transverse Modes

The horizontal transverse modes show the light intensity distributions along the axes parallel to the active layer plane. To control the horizontal transverse modes, gain guiding and index guiding, which were explained in Chapter 3, have been used. In gain guiding, lights propagate only in the optical gain region. In index guiding, lights propagate in a high refractive index region, which is surrounded by low index regions. As shown in Table 5.1, gain guiding is superior to index guiding in fabrication but inferior in lasing characteristics.

Table 5.1. Comparison of gain guiding and index guiding

	Stability	Threshold Current	Fabrication
Gain guiding	Unstable	Large	Simple
Index guiding	Stable	Small	Complicated

(a) Gain Guiding

In gain guiding structures, optical gain regions are formed by restricting the current flowing area. For example, electrodes are selectively evaporated.

Figure 5.26 shows (a) a cross-sectional view of a gain guiding LD seen from a facet, (b) distribution of the carrier concentration, and (c) distribution of the optical gain. The injection current flows from the selectively formed electrode along the arrows by diffusion. As a result, the carrier concentration is largest at the center of the stripe, and it decreases with an increase in the distance from the center. Therefore, the center region of the stripe has the optical gain and the stripe edges have the optical losses.

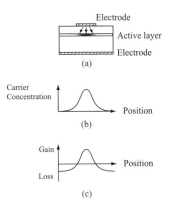

Fig. 5.26. Gain guiding LD: (a) cross-sectional view, (b) distribution of the carrier concentration, and (c) distribution of the optical gain

Here, we consider distribution of the refractive index in the gain guiding structure. With an increase in injection current, the refractive index changes due to the *free carrier plasma effect*, Joule heating in the active layer, and the *spatial hole-burning*.

When the free carriers are induced to vibrate with the frequency of a light, the phases of the free carrier vibrations shift and cancel out the electrical polarizations of the lattice atoms. This phenomenon is referred to as the free carrier plasma effect. A change in the refractive index Δn_{rf} due to the free carrier plasma effect is given by

$$\Delta n_{\mathrm{rf}} = -\frac{e^2}{2m^* \omega^2 \varepsilon_0 n_{\mathrm{r}}}\, n, \qquad (5.78)$$

which is proportional to the injected carrier concentration n. Here, e is the elementary charge, m^* is the effective mass of the free carriers, ω is the angular frequency of a light, ε_0 is permittivity in a vacuum, and n_{r} is the refractive index when the carriers are not injected. On the derivation of (5.78), see Appendix F. When the carrier concentration n is on the order of $10^{18}\,\mathrm{cm}^{-3}$, a decrease in the refractive index is on the order of 10^{-3}. Because the carrier concentration is large at the center of the stripe, the refractive index in the center is lower than that of the surrounding regions according to (5.78). As a result, the light is not completely confined to the active layer and is radiated to the surrounding regions, which is called the *antiguiding effect*.

With an increase in injection current, the active layer is heated by Joule heating. Therefore, the refractive index increases, and this change Δn_{rT} is expressed as

$$\Delta n_{rT} = (2 \sim 5) \times 10^{-4} \Delta T \tag{5.79}$$

where ΔT is an increase in the temperature of the active layer, which is expressed in the units of Kelvin. As opposed to the free carrier plasma effect, the increase in the refractive index due to Joule heating of the active layer contributes to confine the light to the active layer, which is referred to as the *guiding effect*.

When we further increase the injection current and have a large light output, a lot of carriers are consumed due to the stimulated emission. Because stimulated emission efficiently takes places at a large optical gain region, the carrier concentration at the center of the stripe is lower than that of its surrounding regions. This phenomenon is known as the spatial hole-burning, which increases the refractive index at the center region, resulting in a large confinement of the light to the active layer.

As explained earlier, the horizontal distribution of the refractive index is determined by the free carrier plasma effect, Joule heating of the active layer, and the spatial hole-burning. Hence, the horizontal transverse modes show complicated behaviors, such as changes in the positions or multiple peaks, according to a value of the injection current. When the horizontal transverse modes change, *kinks* are observed in *I-L* curves.

In gain guiding structures, wave fronts of the horizontal transverse modes bend convexly in the propagation direction, while those of the vertical transverse modes are close to plane waves. As a result, the *beam waist*, in which the beam diameter is minimum, of the horizontal transverse modes is placed inside the optical cavity, while that of the vertical transverse modes is located on a facet of a semiconductor laser. Such a difference in the positions of the beam waists for the vertical and horizontal transverse modes is called *astigmatism*. When there is astigmatism, we cannot focus both the vertical and horizontal transverse modes on a common plane by a oneaxially symmetric convex lens, and we obtain only defocused images.

(b) Index Guiding

Index guiding structures have an intentionally formed refractive index distribution, as shown in Fig. 5.27. In the index guiding structures, both the horizontal and the vertical transverse modes are close to plane waves, which do not generate astigmatism.

To obtain stable horizontal transverse modes, we need that (1) $\Delta n_r > |\Delta n_{rf}|$ where Δn_r is a difference in the refractive indexes between the active layer and its surrounding layers and Δn_{rf} is a change in the refractive

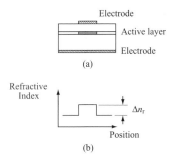

Fig. 5.27. Index guiding LD: (a) cross-sectional view and (b) distribution of the refractive index

index due to the free carrier plasma effect; (2) the cutoff condition for the higher-order modes is satisfied; and (3) the active layer width is shorter than the *diffusion length* of the carriers. Condition (1) is required to achieve index guiding even in a high injection current. Condition (2) is introduced to obtain a single horizontal transverse mode. If there are multiple higher-order horizontal transverse modes, *mode hopping* or the *mode competition* takes place between various modes according to operating conditions, which leads to unstable horizontal transverse modes. To avoid these unstable conditions, only the fundamental horizontal transverse mode should exist in the optical waveguides. Finally, condition (3) is needed to achieve uniform distributions of the carrier concentration in the active layer, which reduces spatial hole-burning. The active layer width is usually about $2\,\mu$m, because the diffusion length of the injected carriers is 2–$3\,\mu$m from the viewpoint of reproducibility. To satisfy conditions (1) and (2), $\Delta n_r/n_r$ is approximately 10^{-2}.

Up to now, many index guiding structures have been developed to efficiently confine both the carriers and the light to the active layer, and the structures are classified into three categories: rib waveguides, ridge waveguides, and buried heterostructures (BHs).

(i) Rib Waveguide

Rib waveguides are the optical waveguides with convex or concave regions, which are suitable for semiconductor lasers whose active layers cannot be exposed to the air by etching. For example, AlGaAs active layers are easily oxidized in the air, and their emission efficiencies drastically decrease.

As an example of a rib waveguide, Fig. 5.28 shows a plano convex waveguide (PCW) structure, in which the semiconductor layers are grown on a preetched substrate. Here, solid line *a* and broken line *b* show the flowing paths of the electric currents. Path *a* is a direction of a forward current across the pn-junction, and the electrical resistance along this path is low. On the other hand, in path *b*, the electric current flows through a *pnpn structure* (*thyristor*).

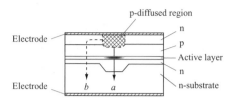

Fig. 5.28. Rib waveguide

Figure 5.29 shows I-V characteristics in the pn and pnpn structures. When the applied voltage is below a switching voltage V_s, a pn-junction inside the pnpn structure is reversely biased, and the electric current hardly flows. Once the applied voltage exceeds V_s, the pnpn structure shows similar I-V characteristics to those of the pn-junctions in forward bias. Therefore, by designing V_s to be larger than the applied voltage across the pn-junction, the electric current hardly flows through path b.

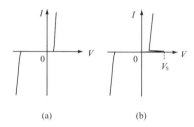

Fig. 5.29. Current versus voltage (I-V) characteristics in (a) pn and (b) pnpn structures

Note that the electric current flowing regions are broad, because a current constriction structure is not formed below the p-diffused region. In other words, rib waveguides are not optimized for confinement of the carriers to the active layers.

(ii) Ridge Waveguide

Ridge waveguides are the optical waveguides with a convex region. Because they are easily fabricated by etching after epitaxial growth, low-cost semiconductor lasers are expected. When the active layer materials cannot be exposed to the air, etching is stopped above the active layer, as shown in Fig. 5.30.

(iii) Buried Heterostructure

Buried heterostructures (BHs), in which the active layer is surrounded by regrown regions, are fabricated as follows: At first, epitaxial layers are grown

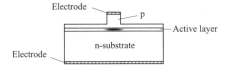

Fig. 5.30. Ridge waveguide

on a semiconductor substrate, which is followed by etching to form a stripe. This stripe is buried by the second epitaxial growth, and the buried regions prevent the injection current from flowing. Although fabrication processes are complicated, low-threshold and high-efficiency operations are obtained because of efficient confinement of the carriers and light to the active layers. However, because the active layers are exposed to the air during fabrication, only the active layers insensitive to oxidization are suitable for buried heterostructures. For example, InGaAsP/InP LDs, which are the light sources of the optical fiber communication systems, frequently adopt buried heterostructures. Figure 5.31 shows an example of a buried heterostructure. To constrict the current flowing region, the surrounding regions of the stripe are pnpn structures. Therefore, the electric current efficiently flows in path *a* but hardly flows along path *b*. Also, a current constriction structure is formed below the p-diffused region, and the carriers are efficiently injected into the active layer.

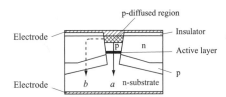

Fig. 5.31. Buried heterostructure (BH)

5.11 Longitudinal Modes

The *longitudinal modes*, or *axial modes*, which determine the resonance wavelengths of the cavity, show the light intensity distributions along the cavity axes. Figure 5.32 shows examples of oscillation spectra for (a) a *multimode operation* and (b) a *single-mode operation*.

Semiconductor lasers use *interband transitions* to obtain the optical gain, and the optical gain spectrum has a width of about 10 nm. Also, the Fabry-Perot cavities have a lot of resonance modes, which leads to low mode selectivity. For these two reasons, the Fabry-Perot LDs tend to oscillate in

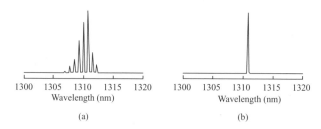

Fig. 5.32. Oscillation spectra of semiconductor lasers: (a) multimode operation and (b) single-mode operation

multimodes. However, we do not need single-mode LDs for applications such as compact discs, laser printers, bar-code readers, laser pointers, or short-haul optical fiber communication systems, in which Fabry-Perot LDs are used.

In long-haul, large-capacity optical fiber communication systems, we need single-mode LDs because the optical fibers have *dispersions* that their refractive indexes depend on the wavelengths and modes of the lights. Due to the dispersions, the propagation speed of light changes according to the wavelengths and modes of the lights. If semiconductor lasers show multimode operations, the optical pulses broaden in a time domain while propagating the optical fibers. With an increase in the transmission distance and a decrease in the pulse spacing, adjacent optical pulses tend to overlap each other. As a result, the receivers cannot resolve sequentially transmitted optical pulses, as shown in Fig. 5.33.

To achieve single-longitudinal-mode operations, the DFB-LDs, the DBR-LDs, the surface emitting LDs, the cleaved coupled cavity (C^3) LDs have been developed, in which the optical cavities select only one lasing mode. These single-longitudinal-mode LDs will be explained in Chapter 6; we focus on the longitudinal modes of Fabry-Perot LDs in this chapter.

Note that the longitudinal modes change with the transverse modes because the effective refractive indexes of the optical waveguides depend on the transverse modes. To achieve single-longitudinal-mode operations, we also have to obtain a single-transverse mode.

Fig. 5.33. Light pulses in (a) a transmitter and (b) a receiver

5.11.1 Static Characteristics of Fabry-Perot LDs

Spacing of the longitudinal modes in the Fabry-Perot LDs is the same as the free spectral range λ_{FSR} in (4.15), which is written as

$$\lambda_{\text{FSR}} = \frac{{\lambda_0}^2}{2n_{\text{r}\lambda}L}, \tag{5.80}$$

where λ_0 is a center wavelength of a light, L is the cavity length, and $n_{\text{r}\lambda}$ is the *equivalent refractive index*, which is the effective refractive index with dispersion, and is given by

$$n_{\text{r}\lambda} = n_{\text{r}} \left(1 - \frac{\lambda_0}{n_{\text{r}}} \frac{\mathrm{d}n_{\text{r}}}{\mathrm{d}\lambda} \right). \tag{5.81}$$

Here, n_{r} is the effective refractive index of the optical waveguide for λ_0. Generally, we have $\mathrm{d}n_{\text{r}}/\mathrm{d}\lambda < 0$, and the equivalent refractive index $n_{\text{r}\lambda}$ is larger than n_{r}. For $\lambda_0 = 1.55\,\mu\text{m}$, $n_{\text{r}\lambda} = 3.5$, and $L = 300\,\mu\text{m}$, λ_{FSR} is $11.4\,\text{Å}$.

Figure 5.34 shows a relationship between the gain spectrum and the longitudinal modes. The Fabry-Perot LDs oscillate at the resonance wavelength, which is closest to the gain peak. As a result, the oscillation wavelength for optical gain A is $\lambda_{0\text{A}}$, and that for optical gain B is $\lambda_{0\text{B}}$.

With an increase in the carrier concentration n, the *refractive index* decreases due to the free carrier plasma effect and increases due to Joule heating of the optical waveguide. Above the threshold, the carrier concentration in the active layer is almost constant. Therefore, above the threshold, the free carrier plasma effect does not change the refractive index, and Joule heating of the active layer enhances the refractive index with an increase in the injection current I. Hence, the resonance wavelength shifts to a longer wavelength with I ($> I_{\text{th}}$) according to (5.5).

With an increase in n, the *optical gain spectrum* shifts to a shorter wavelength due to the band filling effect and to a longer wavelength due to Joule heating of the optical waveguide. The band filling effect is a phenomenon whereby the carriers fill the energy bands from the bottom, and with an increase in the carrier concentration n, the number of carriers with high energy increases. As a result, the optical gain peak shifts to a higher energy (a shorter wavelength) with an increase in n, as shown in Figs. 2.9 and 2.10. Above the threshold, the carrier concentration is almost constant, and the band filling effect is not dominant. Therefore, the gain peak shifts to a longer wavelength due to Joule heating of the optical waveguide with an increase in I.

As described earlier, with an increase in I above I_{th}, both the resonance wavelength and the gain peak shift to longer wavelengths due to Joule heating of the optical waveguide. Therefore, the oscillation wavelength increases with I. Here, it should be noted that the change rate of the resonance wavelength and that of the gain peak with a temperature T_{j} are different from each other.

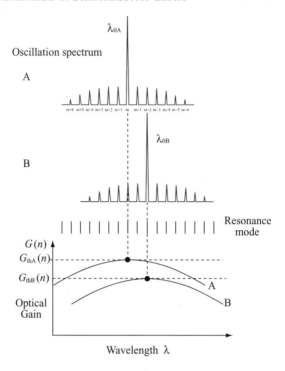

Fig. 5.34. Relationship between the gain spectrum and the longitudinal modes

Hence, with an increase in I or T_j, the oscillation mode jumps, because the resonance wavelength closest to the gain peak at the threshold departs from the gain peak, and other resonance wavelengths approach the gain peak.

Figure 5.35 shows dependence of the longitudinal modes on the injection current I. It is found that the longitudinal modes shift to a longer wavelength, and the mode jumps to the other mode at several values of I. Also, according to an increase or decrease in I, there exist *hysteresis loops*. Similar phenomena are observed when T_j is changed and I is kept constant. The cause of these hysteresis loops is that the optical gain concentrates on the oscillating longitudinal mode, and the optical gains for the other modes are suppressed due to coupling of modes as explained in Section 5.9.

The carriers, which interact in the oscillation mode, are consumed by the radiative recombination due to the stimulated emission. However, the *intraband relaxation time* of the carriers is on the order of 10^{-12}–10^{-13} s, and the other carriers promptly compensate the consumed carriers. Therefore, the number of carriers related to the unoscillating modes decreases, which reduces the optical gains for the unoscillating modes. Figure 5.36 schematically shows how the other carriers compensate the consumed carriers, which exist inside a rectangle.

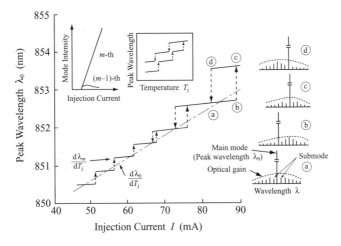

Fig. 5.35. Dependence of longitudinal modes on the injection current

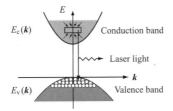

Fig. 5.36. Intraband relaxation

5.11.2 Dynamic Characteristics of Fabry-Perot LDs

(a) Turn-on Delay Time and Relaxation Oscillation

(i) Turn-on Delay Time

We assume that a step pulsed current is injected into a semiconductor laser. As shown in Fig. 5.37 (a), a *bias* current density J_b is below the threshold current density J_{th}, and a pulsed current density J_p is injected to the semiconductor laser at time $t = t_{on} = 0$. Note that the pulse width is much larger than the carrier lifetime τ_n. The carrier concentration n increases from a bias value n_b with a time constant τ_n and reaches the threshold carrier concentration n_{th} at $t = t_d$, as shown in Fig. 5.37 (b). This t_d is called the *turn-on delay time*, and at $t = t_d$ laser oscillation starts. After the start of laser oscillation, the carrier concentration n and the photon density S show *relaxation oscillations*, as shown in Fig. 5.37 (b) and (c).

Let us calculate the turn-on delay time t_d using the rate equations. For simplicity, we neglect coupling of the spontaneous emission to the lasing

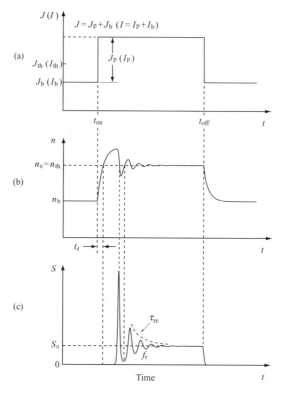

Fig. 5.37. Turn-on delay time and relaxation oscillation: (a) current density, (b) carrier concentration, and (c) photon density

mode. Therefore, when $n < n_{\text{th}}$, the photon density is $S = 0$. As a result, (5.20) reduces to

$$\frac{dn}{dt} = \frac{J}{ed} - \frac{n}{\tau_n}. \tag{5.82}$$

From the assumption, the current density J is given by

$$J = J_{\text{p}} \cdot u(t) + J_{\text{b}},$$

$$u(t) = \begin{cases} 0 & (t < 0), \\ 1 & (t \geq 0). \end{cases} \tag{5.83}$$

Substitution of (5.83) into (5.82) is followed by the *Laplace transform*. If we express a Laplace transform of $n(t)$ as $N(s)$ and let $n(0) = n_{\text{b}} = \tau_n J_{\text{b}}/(ed)$, we obtain

$$sN(s) - n(0) = sN(s) - \frac{\tau_n J_{\text{b}}}{ed} = \frac{J_{\text{p}} + J_{\text{b}}}{ed} \frac{1}{s} - \frac{1}{\tau_n} N(s). \tag{5.84}$$

Hence, $N(s)$ is written as

$$N(s) = \left(\frac{1}{s} - \frac{1}{s + \tau_n^{-1}}\right) \frac{\tau_n(J_p + J_b)}{ed} + \frac{\tau_n J_b}{ed} \frac{1}{s + \tau_n^{-1}}. \tag{5.85}$$

When we take an *inverse Laplace transform* of $N(s)$, we have

$$
\begin{aligned}
n(t) &= \frac{\tau_n(J_p + J_b)}{ed} u(t) - \frac{\tau_n(J_p + J_b)}{ed} e^{-t/\tau_n} + \frac{\tau_n J_b}{ed} e^{-t/\tau_n} \\
&= \frac{\tau_n(J_p + J_b)}{ed} u(t) - \frac{\tau_n J_p}{ed} e^{-t/\tau_n}.
\end{aligned} \tag{5.86}
$$

In $t \geq 0$, we have $u(t) = 1$, and then (5.86) is expressed as

$$n(t) = \frac{\tau_n J}{ed} - \frac{\tau_n J_p}{ed} e^{-t/\tau_n}, \tag{5.87}$$

where

$$J_p + J_b = J. \tag{5.88}$$

At $t = t_d$, the carrier concentration n reaches the threshold carrier concentration n_{th}, and from (5.35) we have

$$n(t_d) = n_{th} = \frac{\tau_n J_{th}}{ed}. \tag{5.89}$$

Using (5.87)–(5.89), the turn-on delay time t_d is obtained as

$$t_d = \tau_n \ln \frac{J - J_b}{J - J_{th}}. \tag{5.90}$$

For $J_b = 0.8 J_{th}$, $J = 1.2 J_{th}$, and $\tau_n = 2.5\,\text{ns}$, we have $t_d = 1.7\,\text{ns}$.

To generate high speed optical signals by modulating the injection current into the semiconductor lasers, the turn-on delay time t_d should be short. From (5.90), it is found that a large bias current density J_b, a low threshold current density J_{th}, and a short carrier lifetime τ_n are suitable for high-speed modulations.

It is also considered that (5.90) shows a relationship between t_d and τ_n. Hence, by measuring t_d as a function of J or J_b, we can obtain the carrier lifetime τ_n from (5.90). This measured result is shown in Fig. 5.38, where the bias current I_b is zero.

(ii) Relaxation Oscillation

Let us calculate a decay coefficient and an oscillation frequency of the relaxation oscillation.

For brevity, by neglecting coupling of the spontaneous emission to the lasing mode, we solve the rate equations (5.20) and (5.21). Because there

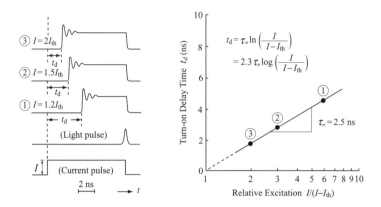

Fig. 5.38. Measurement of carrier lifetime

are no exact analytical solutions, Fig. 5.37 was drawn by numerically analyzing (5.20) and (5.21). However, if we use the *small-signal analysis*, we can
obtain approximate analytical solutions, which clearly give us their physical meanings. To perform the small-signal analysis, we express the carrier
concentration n, the photon density S, and the current density J as

$$n = n_{c0} + \delta n, \quad S = S_0 + \delta S, \quad J = J_0 + \delta J > J_{th},$$
$$n_{c0} \gg \delta n, \quad S_0 \gg \delta S, \quad J_0 \gg \delta J, \tag{5.91}$$

where n_{c0}, S_0, J_0 are the carrier concentration, the photon density, and the
current density in a steady state, respectively, and δn, δS, and δJ are *deviations* from each steady-state value. If we assume $J_b \gg J_p$ and exclude
several initial sharp peaks in the relaxation oscillation, the conditions for the
small-signal analysis are satisfied. Here, we put $J_0 = J_b$ and $\delta J = J_p$.

Neglecting coupling of the spontaneous emission to the lasing mode, (5.20)
and (5.21) reduce to

$$\frac{dn}{dt} = \frac{J}{ed} - G(n)S - \frac{n}{\tau_n}, \tag{5.92}$$

$$\frac{dS}{dt} = G(n)S - \frac{S}{\tau_{ph}}. \tag{5.93}$$

In a steady state ($d/dt = 0$), from (5.91)–(5.93), we have

$$\frac{J_0}{ed} - G(n_{c0})S_0 - \frac{n_{c0}}{\tau_n} = 0,$$
$$G(n_{c0}) = \frac{1}{\tau_{ph}}. \tag{5.94}$$

Substituting (5.91) into (5.22) results in

$$G(n) = G(n_{c0} + \delta n) = \Gamma_a g_0(n_{c0} + \delta n - n_0)$$

$$= \Gamma_a g_0(n_{c0} - n_0) + \Gamma_a g_0 \delta n = G(n_{c0}) + \frac{\partial G}{\partial n} \delta n, \qquad (5.95)$$

where we have introduced the *differential gain*, which is defined as

$$\Gamma_a g_0 \equiv \frac{\partial G}{\partial n}. \qquad (5.96)$$

Inserting (5.91) into (5.92) and(5.93) with the help of (5.94) and (5.95) and then neglecting the second-order small term $\delta n \cdot S$, we have the rate equations on the deviations δn and δS as

$$\frac{d}{dt}\delta n = \frac{\delta J}{ed} - \frac{\delta S}{\tau_{ph}} - \frac{\partial G}{\partial n} S_0 \delta n - \frac{\delta n}{\tau_n}, \qquad (5.97)$$

$$\frac{d}{dt}\delta S = \frac{\partial G}{\partial n} S_0 \delta n. \qquad (5.98)$$

From (5.98), δn is expressed as

$$\delta n = \frac{1}{(\partial G/\partial n)S_0} \frac{d}{dt}\delta S. \qquad (5.99)$$

Substituting (5.97) and (5.99) into a time derivative of (5.98), we have

$$\frac{d^2}{dt^2}\delta S + \left(\frac{\partial G}{\partial n} S_0 + \frac{1}{\tau_n}\right) \frac{d}{dt}\delta S + \frac{\partial G}{\partial n} \frac{S_0}{\tau_{ph}} \delta S = \frac{\partial G}{\partial n} \frac{S_0}{ed} \delta J. \qquad (5.100)$$

This equation shows a relaxation oscillation on the deviation of the photon density δS. As a result, if we write the *decay coefficient* as γ_0 and the *oscillation angular frequency* as ω_r, we obtain

$$\gamma_0 = \frac{\partial G}{\partial n} S_0 + \frac{1}{\tau_n}, \qquad (5.101)$$

$$\omega_r^2 = \frac{\partial G}{\partial n} \frac{S_0}{\tau_{ph}}. \qquad (5.102)$$

From (5.102), the *relaxation oscillation frequency* f_r is given by

$$f_r = \frac{1}{2\pi} \sqrt{\frac{\partial G}{\partial n} \frac{S_0}{\tau_{ph}}}. \qquad (5.103)$$

Note that the decay coefficient γ_0 is related to the *decay time* τ_{re} as

$$\gamma_0 = \frac{1}{\tau_{re}}. \qquad (5.104)$$

To generate high-speed optical signals by modulating the injection currents into the semiconductor lasers, the optical pulses have to quickly return to their steady-state values. Therefore, the decay coefficient γ_0 and the relaxation oscillation frequency f_r have to be large. From (5.101) and (5.103), it is found that a large differential gain $\partial G/\partial n$, a large photon density S_0 in a steady state, a short carrier lifetime τ_n, and a short photon lifetime τ_{ph} are suitable for high-speed modulations.

Here, we consider a relationship between high-speed modulations and other characteristics. As found in (5.40) and (5.96), a large $\partial G/\partial n$ leads to a low J_{th}, while short τ_n and τ_{ph} increase J_{th}. Hence, for simultaneous low-threshold and high-speed operations, we need to obtain a large $\partial G/\partial n$, which is achieved in the quantum well LDs (see Chapter 7). With regard to system applications, if S_0 is large, light intensity is detected by a receiver even when optical pulses are not transmitted. As a result, an *extinction ratio* decreases and a *signal-to-noise (S/N) ratio* degrades. Therefore, S_0 should be limited to satisfy specifications of applications.

The decay coefficient γ_0 and the relaxation oscillation frequency f_r can also be expressed using the current density J. From (5.34), the threshold carrier concentration n_{th} and the transparent carrier concentration n_0 are written as

$$n_{th} = \frac{\tau_n}{ed} J_{th}, \quad n_0 = \frac{\tau_n}{ed} J'_0, \tag{5.105}$$

where J'_0 is the transparent current density, in which a material is transparent. From (5.39), we have

$$n_{th} - n_0 = \frac{1}{\Gamma_a g_0 \tau_{ph}}. \tag{5.106}$$

Substituting (5.96) and (5.105) into (5.106), we obtain

$$\frac{\partial G}{\partial n} = \frac{ed}{\tau_n \tau_{ph} (J_{th} - J'_0)}. \tag{5.107}$$

Inserting (5.42) and (5.107) into (5.101) and (5.103), the decay coefficient γ_0 and the relaxation oscillation frequency f_r are expressed as

$$\gamma_0 = \frac{1}{\tau_n} \frac{J - J'_0}{J_{th} - J'_0}, \tag{5.108}$$

$$f_r = \frac{1}{2\pi} \sqrt{\frac{1}{\tau_n \tau_{ph}} \frac{J - J_{th}}{J_{th} - J'_0}}. \tag{5.109}$$

(b) Relationship between the Relaxation Oscillation and the Longitudinal Modes

As shown in Fig. 5.37 (b), during the relaxation oscillation, the carrier concentration n deviates from n_{th} by about $\pm 10\%$. Due to this modulation in n, the optical gain and the refractive index simultaneously change, which alters the longitudinal modes.

First, we consider the optical gain. When the carrier concentration n is larger than n_{th}, the optical gain g exceeds the threshold gain g_{th} and the longitudinal modes with the optical gain $g \geq g_{\text{th}}$ show laser operations. Therefore, in the beginning of the relaxation oscillation, multimode laser oscillations are observed. With a decay in the relaxation oscillation, the number of the lasing longitudinal modes decreases.

Secondly, we treat the refractive index. The refractive index is modulated by the free carrier plasma effect. When a deviation in the carrier concentration δn is 2×10^{17} cm^{-3}, the positions of the longitudinal modes shift to a shorter wavelength by about 4 Å from steady-state values. Such dynamic changes in the longitudinal modes due to modulations of n are called *chirping*. When the chirping takes place, the linewidth of the time-averaged light output spectra broaden, as shown in Fig. 5.39. Multimode laser operations and chirping are not suitable for long-haul, large-capacity optical fiber communication systems, because the optical fibers have dispersions, as described earlier.

It should be noted that the decay time of the relaxation oscillation is on the order of nanoseconds. Therefore, the effect of Joule heating with time scale longer than microseconds is negligible.

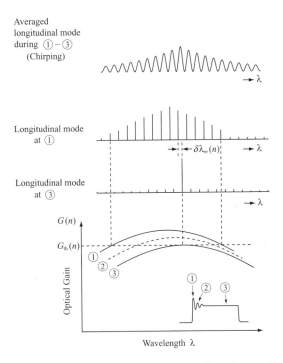

Fig. 5.39. Multimode operation during relaxation oscillation

5.12 Modulation Characteristics

5.12.1 Lightwave Transmission Systems and Modulation

From the viewpoint of light modulations in lightwave transmission systems, there are an *intensity-modulation/direct-detection* system and a *coherent* system. With regard to the modulations of laser beams, there are a *direct modulation* of semiconductor lasers and an *external modulation* using optical modulators.

(a) Intensity-Modulation/Direct-Detection System

Figure 5.40 shows the intensity-modulation/direct-detection system. In this system, a transmitter sends optical signals by modulating the light intensity, while a receiver directly detects changes in the light intensity, and transforms these optical signals into electric signals. This system is simpler and more cost-effective than the coherent one. Therefore, this intensity-modulation/direct-detection system is used in contemporary optical fiber communication systems. A problem with transmission distance, which was inferior to the coherent system, has been solved by the advent of *optical fiber amplifiers*.

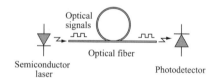

Fig. 5.40. Intensity-modulation/direct-detection system

(b) Coherent System

In coherent systems, there are *amplitude shift keying* (ASK), *frequency shift keying* (FSK), and *phase shift keying* (PSK) systems according to modulations of the laser beam. As shown in Fig. 5.41, the modulated light from a master light source LD_1 and a light from a slave (local) light source LD_2 are simultaneously incident on a receiver. In the receiver, interference of these two laser beams generates an optical beat signal, which is converted to an electric signal.

With an increase in the light output of LD_2, the S/N ratio is improved, which leads to long-haul optical fiber communication systems. However, to obtain the optical beat, we have to prepare two semiconductor lasers, whose laser lights have narrow spectral linewidths, almost common wavelengths, and the same polarizations. Moreover, we need *polarization controllers* to achieve the same polarizations, because the polarization of the lightwave changes with an increase in the propagation distance due to contortion of the laying optical fibers. Also, electronic circuits in the receiver are complicated. As a result, the cost of the coherent system is much higher than that of the intensity-modulation/direct-detection system.

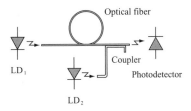

Fig. 5.41. Coherent system

(c) Direct Modulation

In direct modulation of semiconductor lasers, the injection currents into semiconductor lasers are modulated. During direct modulations, multimode operations, chirping, and changes in the turn-on delay time take place. In short-haul optical fiber communication systems, as in a building, these problems are not as serious, and the Fabry-Perot LDs can be used as light sources. In contrast, for long-haul optical fiber communication systems with a transmission distance of more than several tens of kilometers, the Fabry-Perot LDs cannot satisfy the system specifications. Therefore, the DFB-LDs showing stable single-longitudinal-mode operations are used as light sources.

In the coherent system, we would like to modulate only one of the amplitude, frequency, or phase of light. However, they all simultaneously change, when conventional semiconductor lasers are directly modulated. To overcome these problems, semiconductor lasers containing phase-control regions have been developed.

(d) External Modulation

In external modulation of laser beams, the injection currents into semiconductor lasers are kept constant, and the optical modulators modulate laser beams emitted from semiconductor lasers. Because the injection currents into semiconductor lasers are constant, relaxation oscillations do not exist, as opposed to direct modulation. Therefore, we can avoid multimode operations. Also, with a small change in the refractive index in the optical modulators, chirping is low. However, there are problems such as costs and optical coupling efficiencies between the optical modulators and semiconductor lasers. To overcome these problems, integrated light sources of the DFB-LDs and optical modulators have been developed and used in contemporary long-haul, large-capacity optical fiber communication systems.

Optical modulators often use *Franz-Keldysh effect* or *quantum confined Stark effect* (QCSE) as their operating principles. Franz-Keldysh effect is a shift of a fundamental absorption edge to a longer wavelength with an increase in the bias voltage of bulk semiconductors. Quantum confined Stark effect is a shift of an *exciton* absorption peak to a longer wavelength with an increase in the bias voltage of the semiconductor quantum structures.

In the following, we focus on direct modulation of semiconductor lasers, because it is helpful to understand peculiar characteristics of semiconductor lasers.

5.12.2 Direct Modulation

(a) Dependence on Modulation Frequency

The relaxation oscillations of semiconductor lasers are analogous to a *transient response theory* of electric circuits, while dependence on modulation frequency corresponds to an *alternating current theory*.

When a deviation in the current density δJ is related to the steady-state current density J_0 as $|\delta J| \ll J_0$, we can use the small-signal analysis. We assume that a deviation from J_0 is expressed as $\delta J(\omega)e^{i\omega t}$. As a result, a deviation in the carrier concentration δn and that in the photon density δS are expressed as $\delta n = \delta n(\omega)e^{i\omega t}$ and $\delta S = \delta S(\omega)e^{i\omega t}$, respectively. Substituting these into (5.97) and (5.98), we have

$$\delta n(\omega) = -\frac{i\omega}{D(\omega)}\frac{\delta J(\omega)}{ed}, \tag{5.110}$$

$$\delta S(\omega) = -\frac{\tau_{ph}\,\omega_r^{\,2}}{D(\omega)}\frac{\delta J(\omega)}{ed}, \tag{5.111}$$

where

$$D(\omega) = \omega^2 - \omega_r^{\,2} - i\,\omega\gamma_0, \tag{5.112}$$

$$\gamma_0 = \frac{\partial G}{\partial n}S_0 + \frac{1}{\tau_n}, \tag{5.113}$$

$$\omega_r^{\,2} = \frac{\partial G}{\partial n}\frac{S_0}{\tau_{ph}}. \tag{5.114}$$

Note that (5.113) and (5.114) are equal to (5.101) and (5.102), respectively.

If we define a *modulation efficiency* $\delta(\omega)$ as the number of photons generated per injected electron, $\delta(\omega)$ is written as

$$\delta(\omega) = \left| \frac{\delta S(\omega)}{\delta J(\omega)/(ed)} \right| = \frac{\tau_{\text{ph}}\,\omega_{\text{r}}^2}{|D(\omega)|} = \frac{\tau_{\text{ph}}\,\omega_{\text{r}}^2}{\sqrt{(\omega^2 - \omega_{\text{r}}^2)^2 + \omega^2 \gamma_0^2}}, \tag{5.115}$$

where (5.111) and (5.112) were used. From (5.115), we obtain

$$\frac{\delta(\omega)}{\delta(0)} = \frac{\omega_{\text{r}}^2}{\sqrt{(\omega^2 - \omega_{\text{r}}^2)^2 + \omega^2 \gamma_0^2}}. \tag{5.116}$$

As found in (5.116), the modulation efficiency $\delta(\omega)$ shows resonance characteristics, and the *resonance angular frequency* is identical to the relaxation oscillation angular frequency ω_{r}.

Figure 5.42 shows the modulation efficiency $\delta(f)$ as a function of the modulation frequency $f = \omega/2\pi$ with the injection current density J as a parameter. The modulation efficiency $\delta(f)$ has a maximum value at a resonance frequency f_{r}, and $\delta(f)$ drastically decreases with an increase in f over f_{r}. Therefore, the resonance frequency f_{r} indicates the highest limit in the modulation frequency. As shown in (5.109), with an increase in J, f_{r} is enhanced, which results in a large modulation bandwidth. It should be noted that electrical resistance and capacitance of semiconductor lasers also affect the modulation bandwidth.

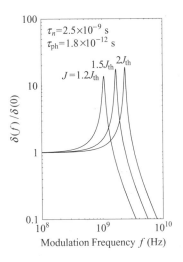

Fig. 5.42. Resonance phenomena

(b) Analog Modulation and Digital Modulation

In the direct modulations of semiconductor lasers, there are *analog modulation* and *digital modulation*, as shown in Fig. 5.43.

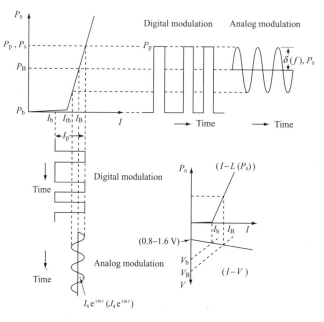

Fig. 5.43. Analog modulation and digital modulation

Analog modulation transforms a change in the injection current into a change in light intensity and requires a high linearity in the I-L curve. The upper limit in the modulation frequency is the resonance frequency f_r, and the causes of high-frequency distortions are nonlinearity in the I-L curve and Joule heating during large amplitude modulations. Typical noises are optical feedback noises induced by the lights reflected from edges of the optical fibers and the modal noises generated in the optical fibers. Compared with digital modulation, analog modulation can transmit more information with lower modulation frequencies, but it is easily affected by distortions of optical signals during transmissions. As a result, analog modulation is used for short-haul, large-capacity optical fiber communication systems such as cable televisions (CATVs).

Digital modulation assigns the signal of 1/0 to ON/OFF of the light intensity. To obtain the optical pulses, which immediately follow the pulsed currents, DC bias currents are injected into semiconductor lasers to shorten the turn-on delay time. Compared with analog modulation, digital modulation is less affected by distortions of optical signals during transmission. As a result, digital modulation is used for long-haul, large-capacity optical fiber communication systems. However, to transmit a lot of information, such as motion pictures, higher modulation frequency than that of analog modulation is required.

In the following, we will explain digital modulation in more depth. According to the original signals 1 and 0, there are *return-to-zero* (RZ) signal and *nonreturn-to-zero* (NRZ) signal, as shown in Fig. 5.44. In the RZ signal, a signal level is returned to zero after each signal is transmitted, which results in a large S/N ratio. However, we need a short pulse width, which leads to a high modulation frequency. In the NRZ signal, a signal level is not returned to zero after each signal is transmitted. Hence, if the original signal 1 continues, the signal level is kept high until the original signal 0 appears. Therefore, the S/N ratio is inferior to that of the RZ signal, but the modulation speed is not as high as that of the RZ signal.

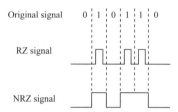

Fig. 5.44. RZ and NRZ signals

Now we consider high-speed modulations to transmit large-capacity information. When the modulation speed is higher than 400 Mb/s, the pulse width is on the order of nanoseconds. In this case, the number of peaks in the relaxation oscillation for each optical pulse is at most two, and a steady state does not exist within each pulse. Hence, a decrease in average light intensity, pattern effect, and multimode operation takes place.

When the relaxation oscillation is sharp, the light intensity pulses have deep valleys and the averaged light intensity decreases, which restricts the transmission distance.

The carrier concentration in the active layer changes according to existence or nonexistence of a preceding optical pulse, which is referred to as the *pattern effect*. With a change in the remaining carrier concentration, the bias level is altered, which modifies the turn-on delay time, as shown in Fig. 5.45. In order to examine the changes in the optical pulses, an optical signal is formed by superimposing the optical pulses with the electrical pulses as the reference. This newly formed optical signal is called the *eye pattern*, because its waveform resembles an eye when a *duty* of the pulse is 50%. When there are large deviations in the turn-on delay time, resolutions in the receiver degrade. In this case, the eye patterns collapse and a lot of jitters appear. When the deviations in the turn-on delay time are small, the eye patterns are

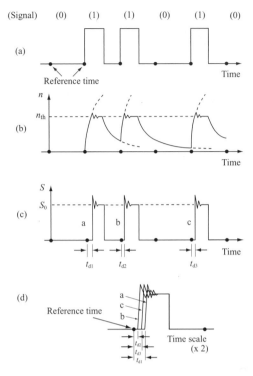

Fig. 5.45. Pattern effect: (a) pulsed injection current, (b) injected carrier concentration, (c) pulsed light intensity, and (d) eye pattern

wide open. Therefore, by observing the eye patterns, we can evaluate quality of the transmitted signals. Figure 5.46 shows examples of the eye patterns for modulation speeds of 4 Gb/s and 2.4 Gb/s with a duty of 50%. In these examples, the eye patterns are clearly open, and high-quality transmission characteristics are obtained.

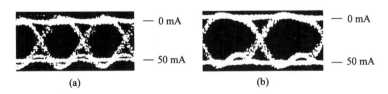

Fig. 5.46. Eye pattern: (a) 4 Gb/s and (b) 2.4 Gb/s

When the relaxation oscillation takes place, the Fabry-Perot LDs show multimode operations, which restrict the transmission distance due to the dispersions in the optical fibers. Here, we describe the dispersions in the optical fibers in detail and evaluate a relationship between the dispersions and the transmission distance.

The optical fibers have *mode dispersion*, *material dispersion*, and *structural dispersion*. Mode dispersion is a change in the effective refractive index of the optical fibers according to a mode, which corresponds to a distribution of light fields. Material dispersion, or *chromatic dispersion*, is a variation in the refractive index of the optical fibers with a wavelength of light. Structural dispersion is a change in the effective refractive index of the optical fibers with a wavelength of light for a common mode; it is caused because the refractive index of the core and that of the cladding of the optical fibers depend on a wavelength of light.

In long-haul, large-capacity optical fiber communication systems, a fundamental mode is used and the material dispersion is dominant. Table 5.2 shows the material dispersions and the optical losses of the glass optical fibers for typical wavelengths of semiconductor lasers.

Table 5.2. Material dispersions and optical losses in glass optical fibers

Wavelength (μm)	Material Dispersion (ps/Å/km)	Loss (dB/km)
0.85	9	$2 - 3$
1.30	~ 0	0.5
1.55	2	< 0.2

If five longitudinal modes lase during the relaxation oscillation, the difference in the longest and shortest wavelengths is approximately 50 Å for semiconductor lasers with a wavelength of 1.55 μm. When this optical signal is transmitted through the optical fibers by 10 km, the lights arrive at a receiver with a time difference of

$$2 \,\text{ps/Å/km} \times 50 \,\text{Å} \times 10 \,\text{km} = 1 \,\text{ns}, \qquad (5.117)$$

and the optical pulse shape degrades.

Here, we summarize key points for long-haul, large-capacity optical fiber communication systems.

For long-haul optical fiber communication systems, we need to use optical fibers with low material dispersions and low optical losses. A wavelength with the lowest dispersion is 1.30 μm, and a wavelength with the lowest optical loss is 1.55 μm. Hence, in the conventional optical fiber communication systems, wavelengths of 1.30 μm and 1.55 μm are used. In Japan, the dispersion-shifted optical fibers whose lowest dispersion wavelength is shifted from 1.30 μm to 1.55 μm are used in trunk-line systems. The light sources are required to show highly stable single-longitudinal-mode operations with low chirping and large average light output. Therefore, the DFB-LDs with excellent single-longitudinal-mode operations are adopted.

For large-capacity optical fiber communication systems, we need to shorten
the turn-on delay time and enlarge the relaxation oscillation frequency (resonance frequency). As a result, DC bias currents are injected into semiconductor lasers. Also, semiconductor laser structures are designed to reduce
capacitance.

Finally, we briefly explain driving circuits of semiconductor lasers. The
threshold current, light output, and oscillation wavelength of semiconductor
lasers change with temperature. Hence, to prevent these changes in characteristics, temperature control circuits are used in the conventional optical
fiber communication systems. Also, to keep the light output constant, a pho-
todiode is placed at a rear facet of each semiconductor laser to monitor the
light output, and *automatic power control* (APC) circuits are adopted.

5.13 Noises

Noises related to semiconductor lasers are as follows.

> ### Noises Peculiar to Semiconductor Lasers
> Quantum noises:
> AM noise (fluctuation in an amplitude of a light)
> FM noise (fluctuation in a frequency of a light)
> Noises on longitudinal modes:
> Mode partition noise
> Mode hopping noise
> ### External Noises
> Optical feedback noise
> Noise due to fluctuations in temperature,
> driving current, and voltage.

5.13.1 Quantum Noises

(a) Fundamental Equations

Amplitude modulating (AM) noise and frequency modulating (FM) noise are
quantum noises. They are caused by the spontaneous emission in a free space
with random amplitudes, frequencies, and phases of the lights. Figure 5.47
shows a relationship between spontaneous emission and the quantum noises
such as AM noise and FM noise. In addition to spontaneous emission, FM
noise is affected by AM noise as follows: Due to AM noise, the amplitude of
the light field fluctuates, which modulates the carrier concentration in the
active layer and leads to *carrier noise*. As a result, through the free carrier plasma effect, the refractive indexes of semiconductors fluctuate, which
results in FM noise. Also, the carrier noise induced by AM noise generates

electric current noise, which changes Joule heating in the active layer. Therefore, the refractive indexes of semiconductors fluctuate, which results in FM noise. AM noise is also altered by the carrier noise induced by AM noise itself.

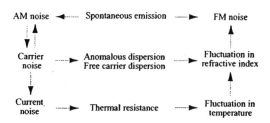

Fig. 5.47. Quantum noises

To analyze the quantum noises, we use *semiclassical theory*, in which electromagnetic fields are treated classically and atomic systems in the fields are considered quantum mechanically [41].

From Maxwell's equations, we have an equation for the electric field $\boldsymbol{E}$ in an optical cavity for semiconductor lasers as

$$\nabla^2 \boldsymbol{E} - \mu\sigma\frac{\partial \boldsymbol{E}}{\partial t} - \mu\varepsilon\frac{\partial^2 \boldsymbol{E}}{\partial t^2} = \mu\frac{\partial^2}{\partial t^2}(\boldsymbol{P} + \boldsymbol{p}), \qquad (5.118)$$

where μ permeability; σ conductivity, which represents the optical losses; ε_0 permittivity in a vacuum; n_r a refractive index of the material ($\varepsilon \equiv \varepsilon_0\, n_\mathrm{r}^2$); $\boldsymbol{P}$ a polarization of a medium contributing to the laser transition; and $\boldsymbol{p}$ a polarization source for the spontaneous emission (Langevin source).

To derive (5.118), a vector formula

$$\nabla \times \nabla \times \boldsymbol{E} = \nabla(\nabla \cdot \boldsymbol{E}) - \nabla^2 \boldsymbol{E} \qquad (5.119)$$

was used, and $\nabla \cdot \boldsymbol{E} = 0$ was assumed.

We also suppose that the electric field $\boldsymbol{E}$ and the polarizations $\boldsymbol{P}$ and $\boldsymbol{p}$ are expressed as

$$\boldsymbol{E} = \mathrm{Re}\left[\sum_m E_m(t)\boldsymbol{e}_m(\boldsymbol{r})\right],$$

$$\boldsymbol{P} = \mathrm{Re}\left[\sum_m P_m(t)\boldsymbol{e}_m(\boldsymbol{r})\right], \qquad (5.120)$$

$$\boldsymbol{p} = \mathrm{Re}\left[\sum_m p_m(t)\boldsymbol{e}_m(\boldsymbol{r})\right].$$

Here, a spatial distribution function $\boldsymbol{e}_m(\boldsymbol{r})$ satisfies both the boundary conditions of the optical cavity and the following relation

$$\nabla^2 e_m(r) + \omega_m{}^2 \mu \varepsilon\, e_m(r) = 0. \tag{5.121}$$

Also, $e_m(r)$ is *orthonormalized* as

$$\int_{\text{all volume}} e_m{}^* \cdot e_n \, \mathrm{d}V = \delta_{mn} V, \tag{5.122}$$

where V is a *mode volume*. Substituting (5.120) into (5.118) with the help of (5.121), taking the inner product with $e_n{}^*(r)$, and using (5.122), we obtain

$$\ddot{E}_n + \frac{1}{\tau_{\text{ph}}} \dot{E}_n + \omega_n{}^2 E_n = -\frac{1}{\varepsilon}(\ddot{P}_n + \ddot{p}_n), \tag{5.123}$$

where a dot and a double dot above E_n, P_n, and p_n indicate a first derivative and a second derivative with respect to a time t, respectively; ω_n is the resonance angular frequency of the nth mode; and $\tau_{\text{ph}} = \varepsilon/\sigma$ is the photon lifetime of the optical cavity. Here, we have also assumed $\mu = \mu_0$, which is usually satisfied in the optical materials.

The polarization of a medium $P_n(t)$ is related to the electric field $E_n(t)$ as

$$P_n(t) = \varepsilon_0 \left[X^{(1)} + X^{(3)} |E_n(t)|^2 \right] E_n(t). \tag{5.124}$$

Here, $X^{(1)}$ and $X^{(3)}$ are expressed as

$$X^{(1)} = \frac{\chi^{(1)}}{V} \int_{\text{medium}} e_n{}^* \cdot e_n \, \mathrm{d}V \approx \chi^{(1)} \frac{l}{L},$$

$$X^{(3)} = \frac{\chi^{(3)}}{V} \int_{\text{medium}} (e_n{}^* \cdot e_n)^2 \, \mathrm{d}V \approx \chi^{(3)} \frac{3l}{2L}, \tag{5.125}$$

where $\chi^{(1)}$ and $\chi^{(3)}$ are the first and third optical susceptibilities, respectively; l is the crystal length; L is the cavity length; and V is the mode volume.

Under the assumption of single-mode operation, we express the electric field $E_n(t)$ and the polarization source for the spontaneous emission (Langevin source) $p_n(t)$ as

$$E_n(t) = [A_0 + \delta(t)]\, \mathrm{e}^{\,\mathrm{i}\,[\omega_m t + \phi(t)]},$$

$$-\frac{1}{\varepsilon} \frac{\partial^2 p_n}{\partial t^2} = \Delta(t)\, \mathrm{e}^{\,\mathrm{i}\,[\omega_m t + \phi(t)]}, \tag{5.126}$$

where A_0 an averaged amplitude of the electric field, $\delta(t)$ a deviation in the amplitude of the electric field from A_0, $\phi(t)$ an instantaneous phase, $\Delta(t)$ a random function representing spontaneous emission, and ω_m an average angular frequency of laser light.

We also assume that $\delta(t)$, $\phi(t)$, and $\Delta(t)$ are slowly varying functions compared with ω_m. For brevity, we neglect the carrier fluctuations. We further suppose that

$$\delta(t) \ll A_0, \quad \langle \delta(t) \rangle = \langle \phi(t) \rangle = \langle \Delta(t) \rangle = 0, \tag{5.127}$$

where $\langle \cdots \rangle$ shows a *time average* or an *ensemble average*.

Substituting (5.124) and (5.126) into (5.123), we obtain

$$2\,\mathrm{i}\,\omega_m \left(\frac{\partial \delta}{\partial t} + \mathrm{i}\,A_0 \frac{\partial \phi}{\partial t} \right) + \frac{3 A_0{}^2 X^{(3)}}{n_r{}^2} \left(2\,\mathrm{i}\,\omega_m \frac{\partial \delta}{\partial t} - \omega_m{}^2 \delta \right)$$

$$+ \left[\omega_n{}^2 - \omega_m{}^2 + \mathrm{i}\,\frac{\omega_m}{\tau_{\mathrm{ph}}} - (X^{(1)} + A_0{}^2 X^{(3)}) \frac{\omega_m{}^2}{n_r{}^2} \right] A_0 = \Delta(t), \tag{5.128}$$

where $\varepsilon \equiv \varepsilon_0 \, n_r{}^2$ (n_r: a refractive index of the material) was used.

A real part and an imaginary part of the steady-state solution of (5.128) are given by

$$\omega_m{}^2 = \omega_n{}^2 \left[1 + \frac{X_r^{(1)} + A_0{}^2 X_r^{(3)}}{n_r{}^2} \right]^{-1} \quad : \text{ real,} \tag{5.129}$$

$$A_0{}^2 = -\frac{1}{X_i^{(3)}} \left(X_i^{(1)} + \frac{n_r{}^2}{\omega_m \tau_{\mathrm{ph}}} \right) \quad : \text{ imaginary,} \tag{5.130}$$

where $X^{(1,3)} = X_r^{(1,3)} - \mathrm{i}\,X_i^{(1,3)}$.

Therefore, (5.128) reduces to

$$A_0 \frac{\partial \phi}{\partial t} - \frac{3 A_0{}^2 X_i^{(3)}}{n_r{}^2} \frac{\partial \delta}{\partial t} + \frac{3 A_0{}^2 \omega_m X_r^{(3)}}{2 n_r{}^2} \delta = -\frac{\Delta_r(t)}{2 \omega_m} \quad : \text{ real,} \tag{5.131}$$

$$\left(1 + \frac{3 A_0{}^2 X_r^{(3)}}{n_r{}^2} \right) \frac{\partial \delta}{\partial t} + \frac{3 A_0{}^2 \omega_m X_i^{(3)}}{2 n_r{}^2} \delta = \frac{\Delta_i(t)}{2 \omega_m} \quad : \text{ imaginary,} \tag{5.132}$$

where $\Delta(t) = \Delta_r(t) + \mathrm{i}\,\Delta_i(t)$. These equations are the fundamental equations for the noises in the laser light, when the carrier fluctuations are neglected. In (5.131), the second term on the left-hand side is usually neglected because it is much smaller than the other terms. As a result, the amplitude fluctuation δ and the phase fluctuation ϕ are related to each other by the term including $X_r^{(3)}$.

(b) Spectra of a Laser Light

Among the spectra of laser lights, there are a *power fluctuation spectrum* expressing the AM noise, a *frequency fluctuation spectrum* indicating the FM noise, and a *field spectrum*. Figure 5.48 shows a measurement system

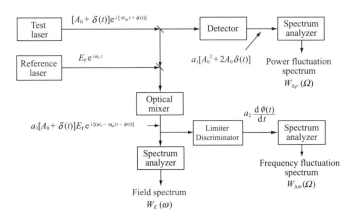

Fig. 5.48. Measurement system for spectra of a laser light

for these spectra. It should be noted that a *spectral linewidth* of laser light usually represents a linewidth of the field spectrum.

To obtain the spectra of laser lights, we first calculate the *autocorrelation functions* of the fluctuations using (5.131) and (5.132). Then, with the help of *Wiener-Khintchine theorem*, we have *spectral density functions*.

(i) Autocorrelation Function of the Amplitude Fluctuation $\delta(t)$

For a usual laser field, $A_0^2 X_r^{(3)}/n_r^2 \ll 1$ is satisfied. As a result, (5.132) reduces to

$$\frac{\partial \delta}{\partial t} + \omega_1 \delta = \frac{\Delta_i(t)}{2\omega_m}, \quad \omega_1 = \frac{3A_0^2 \omega_m X_i^{(3)}}{2n_r^2} > 0. \tag{5.133}$$

Taking a Laplace transform of (5.133), we have

$$-\delta(0) + s\tilde{\delta}(s) + \omega_1 \tilde{\delta}(s) = \frac{\tilde{\Delta}_i(s)}{2\omega_m},$$

$$\tilde{\delta}(s) \equiv \int_0^\infty \delta(t)\, e^{-st}\, dt, \tag{5.134}$$

$$\tilde{\Delta}_i(s) \equiv \int_0^\infty \Delta_i(t)\, e^{-st}\, dt.$$

Supposing $\delta(0) = 0$, (5.134) results in

$$\tilde{\delta}(s) = \frac{\tilde{\Delta}_i(s)}{2\omega_m(s + \omega_1)}. \tag{5.135}$$

Taking an inverse Laplace transform of (5.135), we obtain $\delta(t)$ as

$$\delta(t) = \frac{1}{2\omega_m} \int_0^t \Delta_i(\lambda) e^{-\omega_1(t-\lambda)} \, d\lambda. \tag{5.136}$$

From (5.136), the autocorrelation function $\langle \delta(t+\tau)\delta(t) \rangle$ of the amplitude fluctuation $\delta(t)$ is given by

$$\langle \delta(t+\tau)\delta(t) \rangle = \frac{1}{4\omega_m^2} \int_0^{t+\tau} d\lambda_1 \int_0^t d\lambda_2 \langle \Delta_i(\lambda_1)\Delta_i(\lambda_2) \rangle$$
$$\times e^{-\omega_1(t+\tau-\lambda_1)} e^{-\omega_1(t-\lambda_2)}. \tag{5.137}$$

Here, we assume that the correlation functions of the Langevin sources are written as

$$\langle \Delta_i(\lambda_1)\Delta_i(\lambda_2) \rangle = \langle \Delta_r(\lambda_1)\Delta_r(\lambda_2) \rangle = W \cdot D(\lambda_1 - \lambda_2),$$
$$\langle \Delta_i(\lambda_1)\Delta_r(\lambda_2) \rangle = \langle \Delta_r(\lambda_1)\Delta_i(\lambda_2) \rangle = 0, \tag{5.138}$$

where $D(x)$ is a δ *function* and W is a coefficient related to the spontaneous emission, which will be calculated later. Substituting (5.138) into (5.137), we obtain

$$\langle \delta(t+\tau)\delta(t) \rangle = \frac{W}{8\omega_m^2\omega_1} e^{-\omega_1|\tau|}(1 - e^{-2\omega_1 t}). \tag{5.139}$$

As a result, in a steady state where t is long enough, we have

$$\langle \delta(t+\tau)\delta(t) \rangle = \frac{W}{8\omega_m^2\omega_1} e^{-\omega_1|\tau|}. \tag{5.140}$$

(ii) Autocorrelation Function of the Phase Fluctuation $\phi(t)$

The second term on the left-hand side of (5.131) is small enough compared with the other terms. Therefore, (5.131) reduces to

$$A_0 \frac{\partial \phi}{\partial t} + \frac{3A_0^2\omega_m X_r^{(3)}}{2n_r^2} \delta = -\frac{\Delta_r(t)}{2\omega_m}. \tag{5.141}$$

When $\phi(0) = 0$, integration of (5.141) with respect to t results in

$$\phi(t) = -\frac{3}{4} \frac{X_r^{(3)} A_0}{n_r^2\omega_1} \left[\int_0^t \Delta_i(\lambda) \, d\lambda - \int_0^t \Delta_i(\lambda) e^{-\omega_1(t-\lambda)} \, d\lambda \right]$$
$$-\frac{1}{2A_0\omega_m} \int_0^t \Delta_r(\lambda) \, d\lambda, \tag{5.142}$$

where (5.136) was used.

From (5.133), (5.138), and (5.142), the autocorrelation function $\langle \phi(t_1)\phi(t_2) \rangle$ of the phase fluctuation $\phi(t)$ is given by

$$\langle \phi(t_1)\phi(t_2)\rangle = \frac{W}{4\omega_m{}^2 A_0{}^2}\,(1+\alpha^2) \times \begin{cases} t_1 & (t_1 < t_2), \\ t_2 & (t_1 > t_2). \end{cases} \qquad (5.143)$$

Here, we have introduced

$$\alpha \equiv \frac{X_{\rm r}^{(3)}}{X_{\rm i}^{(3)}}, \qquad (5.144)$$

which is called the α *parameter* or the *spectral linewidth enhancement factor*.
    Using the following relation

$$\begin{aligned}
X &= X_{\rm r} - {\rm i}\,X_{\rm i}, \\
X_{\rm r} &= X_{\rm r}^{(1)} + X_{\rm r}^{(3)}|E_n|^2, \\
X_{\rm i} &= X_{\rm i}^{(1)} + X_{\rm i}^{(3)}|E_n|^2,
\end{aligned} \qquad (5.145)$$

we can rewrite (5.144) as

$$\alpha = \frac{X_{\rm r}^{(3)}}{X_{\rm i}^{(3)}} = \frac{\partial X_{\rm r}}{\partial |E_n|^2}\left(\frac{\partial X_{\rm i}}{\partial |E_n|^2}\right)^{-1} = \frac{\partial X_{\rm r}}{\partial n}\left(\frac{\partial X_{\rm i}}{\partial n}\right)^{-1}, \qquad (5.146)$$

where we have assumed that the light intensity, which is linearly proportional
to $|E_n|^2$, is in proportion to the carrier concentration n. It should be noted
that this α parameter is important to characterize the spectral linewidth and
the optical feedback noise in the semiconductor lasers.

(iii) Autocorrelation Function of the Angular Frequency Fluctuation $\Delta\omega(t)$

Using the averaged angular frequency ω_m and the phase fluctuation $\phi(t)$, we
can express the instantaneous angular frequency $\omega(t)$ of a laser light as

$$\omega(t) = \omega_m + \frac{\partial \phi}{\partial t}. \qquad (5.147)$$

As a result, the fluctuation in the angular frequency $\Delta\omega(t)$ is written as

$$\Delta\omega(t) \equiv \omega(t) - \omega_m = \frac{\partial \phi}{\partial t}. \qquad (5.148)$$

Hence, the autocorrelation function $\langle \Delta\omega(t_1)\Delta\omega(t_2)\rangle$ of the angular frequency
fluctuation $\Delta\omega(t)$ is obtained as

$$\begin{aligned}
\langle \Delta\omega(t_1)\Delta\omega(t_2)\rangle &= \langle \dot{\phi}(t_1)\dot{\phi}(t_2)\rangle \\
&= \frac{W}{4\omega_m{}^2 A_0{}^2}\left[D(t_1 - t_2) + \frac{\alpha^2}{2}\omega_1 {\rm e}^{-\omega_1|t_1 - t_2|}\right],
\end{aligned} \qquad (5.149)$$

where (5.140), (5.141), and (5.146) were used, and a dot above ϕ represents
a first derivative with respect to time.

(iv) Spectral Density Function

To obtain spectra of laser lights, we calculate the spectral density functions of the amplitude fluctuation $\delta(t)$ and the angular frequency fluctuation $\Delta\omega(t)$ using Wiener-Khintchine theorem.

From (5.140), the spectral density function $W_\delta(\Omega)$ of the amplitude fluctuation $\delta(t)$ is obtained as

$$W_\delta(\Omega) = \frac{1}{\pi} \int_{-\infty}^{\infty} \langle \delta(t+\tau)\delta(t) \rangle \, e^{-i\Omega\tau} \, d\tau$$

$$= \frac{W}{4\pi\omega_m{}^2(\Omega^2 + \omega_1{}^2)}. \tag{5.150}$$

From (5.149), the spectral density function $W_{\Delta\omega}(\Omega)$ of the angular frequency fluctuation $\Delta\omega(t)$ is given by

$$W_{\Delta\omega}(\Omega) = \frac{1}{\pi} \int_{-\infty}^{\infty} \langle \Delta\omega(t+\tau)\Delta\omega(t) \rangle \, e^{-i\Omega\tau} \, d\tau$$

$$= \frac{W}{4\pi\omega_m{}^2 A_0{}^2} \left(1 + \frac{\alpha^2\omega_1{}^2}{\Omega^2 + \omega_1{}^2} \right). \tag{5.151}$$

(v) Coefficient W

The generation rate of the spontaneous emission is expressed as $E_{\mathrm{cv}}\hbar\omega_m$ where E_{cv} is the number of photons spontaneously emitted to an oscillation mode per time. The dissipation rate of the spontaneous emission is given by $\Psi_{\mathrm{s}}/\tau_{\mathrm{ph}}$ where Ψ_{s} is an energy of the spontaneous emission coupled to the oscillation mode and τ_{ph} is the photon lifetime. In a steady state, the generation rate is balanced with the dissipation rate. As a result, we have

$$\Psi_{\mathrm{s}} = E_{\mathrm{cv}}\hbar\omega_m\tau_{\mathrm{ph}}. \tag{5.152}$$

The electric field of the spontaneous emission is given by a solution of the following equation

$$\ddot{E}_n + \frac{1}{\tau_{\mathrm{ph}}}\dot{E}_n + \omega_n{}^2 E_n = [\Delta_{\mathrm{r}}(t) + i\,\Delta_{\mathrm{i}}(t)]\,e^{i\omega_m t}, \tag{5.153}$$

where the polarization due to the stimulated emission in (5.123) is neglected and $\omega_n \approx \omega_m$ is assumed.

If we suppose $E(0) = \dot{E}(0)$, as a solution of (5.153), we have

$$E(t) = \frac{1}{\omega_m} \int_0^t d\tau [\Delta_{\mathrm{r}}(\tau) + i\,\Delta_{\mathrm{i}}(\tau)]\,e^{i\omega_m\tau}\,e^{-(t-\tau)/(2\tau_{\mathrm{ph}})}\,\sin[\omega_m(t-\tau)].$$

$$\tag{5.154}$$

Hence, $\langle E^*(t)E(t)\rangle$ is given by

$$
\begin{aligned}
\langle E^*(t)E(t)\rangle &= \frac{1}{\omega_m{}^2} \int_0^t \mathrm{d}\lambda_1 \int_0^t \mathrm{d}\lambda_2\, 2WD(\lambda_1 - \lambda_2)\, \mathrm{e}^{-t/\tau_{\mathrm{ph}}}\, \mathrm{e}^{(\lambda_1+\lambda_2)/(2\tau_{\mathrm{ph}})} \\
&\quad \times\ \sin[\omega_m(t-\lambda_1)]\sin[\omega_m(t-\lambda_2)] \\
&= \frac{W\tau_{\mathrm{ph}}}{\omega_m{}^2},
\end{aligned}
\tag{5.155}
$$

where (5.138) was used.

With the help of (5.155), a steady-state spontaneous emission energy Ψ_{s} is expressed as

$$
\Psi_{\mathrm{s}} = \varepsilon V \langle E^*(t)E(t)\rangle = \frac{\varepsilon V W \tau_{\mathrm{ph}}}{\omega_m{}^2},
\tag{5.156}
$$

$$
V \equiv \int |e_n(\boldsymbol{r})|^2\, \mathrm{d}V,
$$

where ε is the dielectric constant of a material and V is the mode volume.

From (5.152) and (5.156), W is obtained as

$$
W = \frac{\hbar\omega_m{}^3 E_{\mathrm{cv}}}{\varepsilon V}.
\tag{5.157}
$$

(vi) Intensity Fluctuation Spectrum (AM Noise)

To obtain the *intensity fluctuation spectrum*, we first clarify a relationship between the amplitude fluctuation $\delta(t)$ and the intensity fluctuation.

The light emission rate γ from the optical cavity through the mirrors is given by

$$
\gamma = \frac{c}{n_{\mathrm{r}}} \frac{1}{L} \ln \frac{1}{R},
\tag{5.158}
$$

where c is the speed of light in a vacuum, n_{r} is the refractive index of a material, L is the cavity length, and R is the power reflectivity of a facet where both facets are assumed to have a common reflectivity.

Let the electric field of the laser light be E_n; then the light output P is expressed as

$$
P = \varepsilon E_n{}^2 V \gamma.
\tag{5.159}
$$

From (5.126) and (5.159), an averaged light output P_0 and the fluctuation in the light output $\Delta P = P - P_0$ are written as

$$
P_0 = \varepsilon A_0{}^2 V \gamma, \quad \Delta P = 2\varepsilon A_0 \delta V \gamma.
\tag{5.160}
$$

From (5.150), (5.157), (5.158), and (5.160), the spectral density function $W_{\Delta P}(\Omega)$ of the light output fluctuation ΔP is obtained as

$$
\begin{aligned}
W_{\Delta P}(\Omega) &= \frac{1}{\pi} \int_{-\infty}^{\infty} \langle \Delta P(t+\tau)\Delta P(t) \rangle \, \mathrm{e}^{-\mathrm{i}\Omega\tau} \, \mathrm{d}\tau \\
&= 4\varepsilon^2 A_0{}^2 V^2 \gamma^2 W_\delta(\Omega) = \frac{\hbar\omega_m \varepsilon A_0{}^2 V E_{\mathrm{cv}} \gamma^2}{\pi(\Omega^2 + \omega_1{}^2)} \\
&= \frac{\hbar\omega_m E_{\mathrm{cv}} P_0 \gamma}{\pi(\Omega^2 + \omega_1{}^2)} = \frac{\hbar\omega_m c E_{\mathrm{cv}} P_0 \ln(1/R)}{\pi(\Omega^2 + \omega_1{}^2) n_r L}.
\end{aligned}
\tag{5.161}
$$

(vii) Frequency Fluctuation Spectrum (FM Noise)

From (5.151), (5.157), (5.158), and (5.160), the spectral density function $W_{\Delta\omega}(\Omega)$ of the angular frequency fluctuation $\Delta\omega(t)$ is given by

$$
\begin{aligned}
W_{\Delta\omega}(\Omega) &= \frac{\hbar\omega_m E_{\mathrm{cv}}}{4\pi\varepsilon V A_0{}^2} \left(1 + \frac{\alpha^2 \omega_1{}^2}{\Omega^2 + \omega_1{}^2} \right) \\
&= \frac{\hbar\omega_m c E_{\mathrm{cv}} \ln(1/R)}{4\pi P_0 n_r L} \left(1 + \frac{\alpha^2 \omega_1{}^2}{\Omega^2 + \omega_1{}^2} \right).
\end{aligned}
\tag{5.162}
$$

(viii) Field Spectrum

The field spectrum is most frequently used as the laser light spectrum. Contribution of the amplitude fluctuation to the correlation function of the electric field is negligibly small, and the correlation function $\langle E(t+\tau)E(t) \rangle$ is given by

$$
\begin{aligned}
\langle E(t+\tau)E(t) \rangle &= \frac{1}{4} \langle [E(t+\tau) + E^*(t+\tau)][E(t) + E^*(t)] \rangle \\
&= \frac{A_0{}^2}{4} [\mathrm{e}^{-\mathrm{i}\omega_m \tau} \langle \mathrm{e}^{\mathrm{i}\Delta\phi} \rangle + \mathrm{c.c.}],
\end{aligned}
\tag{5.163}
$$

$$
\Delta\phi \equiv \phi(t+\tau) - \phi(t).
$$

The phase fluctuation $\Delta\phi$, which is important to determine the correlation function $\langle E(t+\tau)E(t) \rangle$, is related to many independent spontaneous emission processes. As a result, the distribution of $\Delta\phi$ is given by the *Gaussian distribution function* $g(\Delta\phi)$, which is defined as

$$
g(\Delta\phi) = \frac{1}{\sqrt{2\pi \langle (\Delta\phi)^2 \rangle}} \, \mathrm{e}^{-\frac{(\Delta\phi)^2}{2\langle (\Delta\phi)^2 \rangle}}.
\tag{5.164}
$$

Therefore, we obtain

$$\langle e^{i\Delta\phi}\rangle = \int_{-\infty}^{\infty} g(\Delta\phi)\, e^{i\Delta\phi}\, d(\Delta\phi) = e^{-\frac{1}{2}\langle(\Delta\phi)^2\rangle}. \tag{5.165}$$

Because $(\Delta\phi)^2$ on the right-hand side of (5.165) is independent of a time t, we obtain

$$\langle(\Delta\phi)^2\rangle = \frac{W}{4\,\omega_m{}^2 A_0{}^2}\,(1+\alpha^2)|\tau|, \tag{5.166}$$

where we have assumed $t_1 = t_2 = \tau$ in (5.143). Substituting (5.165) and (5.166) into (5.163) results in

$$\langle E(t+\tau)E(t)\rangle = \frac{A_0{}^2}{2}\,\exp\left[-\frac{W}{8\omega_m{}^2 A_0{}^2}(1+\alpha^2)|\tau|\right]\cos(\omega_m\tau). \tag{5.167}$$

With the help of Wiener-Khintchine theorem, the spectral density function $W_E(\omega)$ of the electric field is given by

$$W_E(\omega) = \frac{1}{\pi}\int_{-\infty}^{\infty}\langle E(t+\tau)E(t)\rangle\, e^{-i\omega\tau}\, d\tau$$

$$\approx \frac{A_0{}^2}{4\pi}\frac{\Delta\omega_0}{(\omega-\omega_m)^2 + (\Delta\omega_0/2)^2}, \tag{5.168}$$

$$\Delta\omega_0 \equiv \frac{\hbar\omega_m c E_{cv}\ln(1/R)}{4 P_0 n_r L}\,(1+\alpha^2). \tag{5.169}$$

From (5.168), the field spectrum is *Lorentzian*, and the FWHM of the field spectrum $\Delta\omega_0$ in (5.169), which is called *modified Schawlow-Townes linewidth formula*, gives the spectral linewidth for semiconductor lasers. In solid-state or gas lasers, the term α^2 can be neglected because $\alpha^2 \ll 1$. In contrast, in semiconductor lasers, the term α^2 is needed because $\alpha^2 > 1$. It should be noted that the fluctuation in the carrier concentration was neglected in the derivation of (5.168) and (5.169).

(c) Spectral Linewidth Enhancement Factor α

The spectral linewidth enhancement factor α is an important parameter that differentiates semiconductor lasers from other lasers. As shown in (5.146), α is given by

$$\alpha = \frac{\partial X_r}{\partial n}\left(\frac{\partial X_i}{\partial n}\right)^{-1}. \tag{5.170}$$

Here, let us express α using the refractive index and the optical gain. Using the complex refractive index $\tilde{n} = n_r - i\kappa$, the complex dielectric constant $\tilde{\varepsilon}$ is written as

$$\tilde{\varepsilon} = \varepsilon_0 \tilde{n}^2 = \varepsilon_0 [(n_r^2 - \kappa^2) - i\, 2n_r\kappa] = \varepsilon_0 (X_r - i\, X_i), \tag{5.171}$$

where

$$X_r = n_r^2 - \kappa^2, \quad X_i = 2n_r\kappa. \tag{5.172}$$

In the vicinity of the bandgap in semiconductors $n_r \gg \kappa$ is satisfied, and we obtain

$$\frac{\partial X_r}{\partial n} = 2n_r \frac{\partial n_r}{\partial n} - 2\kappa \frac{\partial \kappa}{\partial n} \approx 2n_r \frac{\partial n_r}{\partial n}, \tag{5.173}$$

$$\frac{\partial X_i}{\partial n} = 2n_r \frac{\partial \kappa}{\partial n} + 2\kappa \frac{\partial n_r}{\partial n} \approx 2n_r \frac{\partial \kappa}{\partial n}. \tag{5.174}$$

Substituting (5.173) and (5.174) into (5.170) leads to

$$\alpha = \frac{\partial n_r}{\partial n} \left(\frac{\partial \kappa}{\partial n} \right)^{-1}. \tag{5.175}$$

Using the extinction coefficient κ in (5.171), we can express the optical power gain coefficient g as

$$g = -\frac{2\omega_m}{c} \kappa. \tag{5.176}$$

Also, g is related to $G(n)$ in the rate equations as

$$G(n) = \frac{c}{n_r} g. \tag{5.177}$$

Using (5.176) and (5.177), we have

$$\frac{\partial \kappa}{\partial n} = -\frac{n_r}{2\omega_m} \frac{\partial G}{\partial n}. \tag{5.178}$$

Substituting (5.178) into (5.175) results in

$$\alpha = -\frac{2\omega_m}{n_r} \frac{\partial n_r}{\partial n} \left(\frac{\partial G}{\partial n} \right)^{-1}. \tag{5.179}$$

When the optical gain increases with carrier injection ($\partial G/\partial n > 0$) and Joule heating of the active layer is negligibly small, the refractive index decreases ($\partial n_r/\partial n < 0$) due to the free carrier plasma effect. Hence, α defined in (5.179) is positive, and the measured values are between 1 and 7. From (5.179), it is found that we should increase $\partial G/\partial n$ to reduce the value of α, which is achieved in the quantum well LDs explained in Chapter 7.

(d) Suppression of Quantum Noises

The quantum noises derived earlier are summarized as

AM Noise (Intensity Fluctuation Spectrum)

$$W_{\Delta P}(\Omega) = \frac{\omega_m c E_{cv} P_0 \ln(1/R)}{\pi(\Omega^2 + \omega_1^2) n_r L},$$ (5.180)

FM Noise (Frequency Fluctuation Spectrum)

$$W_{\Delta\omega}(\Omega) = \frac{\hbar\omega_m c E_{cv} \ln(1/R)}{4\pi P_0 n_r L}\left(1 + \frac{\alpha^2 \omega_1^2}{\Omega^2 + \omega_1^2}\right),$$ (5.181)

Spectral Linewidth (of a Field Spectrum)

$$\Delta\omega_0 = \frac{\hbar\omega_m c E_{cv} \ln(1/R)}{4 P_0 n_r L}(1 + \alpha^2).$$ (5.182)

From (5.180)–(5.182), to suppress all the quantum noises, we should reduce the mirror loss $(1/L)\ln(1/R)$, for which long cavity length L and large power reflectivity R are needed. However, the long cavities of the *external cavity lasers* are less stable than the cavities of ordinary semiconductor lasers. With a decrease in the spectral linewidth enhancement factor α, which is achieved in the quantum well LDs, the FM noise and spectral linewidth are simultaneously reduced. To suppress the quantum noises, it is also important to stabilize driving circuits and environmental temperature.

5.13.2 Relative Intensity Noise (RIN)

When semiconductor lasers show two-mode operations, the rate equations on the photon densities S_1 and S_2 and the carrier concentration n are written as

$$\frac{dS_1}{dt} = (\alpha_1 - \beta_1 S_1 - \theta_{12} S_2) S_1 + \beta_{sp1}\frac{n}{\tau_n} + F_1(t),$$ (5.183)

$$\frac{dS_2}{dt} = (\alpha_2 - \beta_2 S_2 - \theta_{21} S_1) S_2 + \beta_{sp2}\frac{n}{\tau_n} + F_2(t),$$ (5.184)

$$\frac{dn}{dt} = \frac{I}{eV_A} - [G_1(n) - \beta_1 S_1 - \theta_{12} S_2] S_1$$
$$- [G_2(n) - \beta_2 S_2 - \theta_{21} S_1] S_2 - \frac{n}{\tau_n} + F_n(t).$$ (5.185)

Here, $\alpha_i \equiv G_i(n) - 1/\tau_{phi}$ is the net amplification rate, $G_i(n)$ is the amplification rate, τ_{phi} is the photon lifetime, β_i is the self-saturation coefficient, θ_{ij}

is the cross-saturation coefficient, β_{spi} is the spontaneous emission coupling factor $(i, j = 1, 2)$, τ_n is the carrier lifetime, I is the injection current, e is the elementary charge, and V_A is the volume of the active layer. Fluctuations are expressed by the Langevin noise sources F_1, F_2, and F_n.

The amplification rate $G_i(n)$ is given by

$$G_i(n) = \left[\frac{\partial G_i}{\partial n} \right]_{n=n_{\mathrm{thi}}} (n - n_{0i}), \qquad (5.186)$$

where n_{thi} is the threshold carrier concentration and n_{0i} is the transparent carrier concentration.

In a steady state with weak coupling $(C < 1)$, there are a free-running condition where two modes exist simultaneously and a mode-inhibition condition where only one mode exists. In the neutral and strong couplings, there exists only a mode-inhibition condition, where an oscillation mode suppresses the other mode.

As described earlier, stable simultaneous two-mode operations are obtained only in the weak coupling. In the following calculation,

$$\beta_1 = \beta_2 = \beta, \quad \theta_{12} = \theta_{21} = \frac{2}{3}\beta \qquad (5.187)$$

is used as an example. Other typical physical parameters in the semiconductor lasers are as follows: $[\partial G_i/\partial n]_{n=n_{\mathrm{thi}}} = 10^{-6}\,\mathrm{cm}^3\,\mathrm{s}^{-1}$, $n_{0i} = 6.0 \times 10^{17}\,\mathrm{cm}^3$, $\beta_{\mathrm{spi}} = 10^{-5}$, $\tau_n = 10^{-9}\,\mathrm{s}$, and $\tau_{\mathrm{phi}} = 10^{-12}\,\mathrm{s}$. It is supposed that the injection current I is kept constant but the carrier concentration n fluctuates.

5.13.3 RIN with No Carrier Fluctuations

We express the photon densities S_1 and S_2 as

$$S_1 = S_{10} + \delta S_1(t), \quad S_2 = S_{20} + \delta S_2(t), \qquad (5.188)$$

where S_{10} and S_{20} are the average photon densities of mode 1 and mode 2 in a steady state, respectively; and δS_1 and δS_2 are the fluctuations of mode 1 and mode 2, respectively.

When the Fourier transform of $\delta S_i(t)$ is written as $\delta \tilde{S}_i(\omega)$, the self-correlation functions $\langle \delta \tilde{S}_1(\omega)\delta \tilde{S}_1^*(\omega') \rangle$ and $\langle \delta \tilde{S}_2(\omega)\delta \tilde{S}_2^*(\omega') \rangle$ are given by

$$\langle \delta \tilde{S}_1(\omega)\delta \tilde{S}_1^*(\omega') \rangle = \tilde{S}_{\delta S_1}(\omega) \cdot 2\pi\delta(\omega - \omega'), \qquad (5.189)$$

$$\langle \delta \tilde{S}_2(\omega)\delta \tilde{S}_2^*(\omega') \rangle = \tilde{S}_{\delta S_2}(\omega) \cdot 2\pi\delta(\omega - \omega'). \qquad (5.190)$$

Using (5.189) and (5.190), we define the *relative intensity noises* (RINs) per unit bandwidth as

$$\mathrm{RIN}_1 = \frac{2\tilde{S}_{\delta S_1}(\omega)}{S_{10}{}^2}, \tag{5.191}$$

$$\mathrm{RIN}_2 = \frac{2\tilde{S}_{\delta S_2}(\omega)}{S_{20}{}^2}. \tag{5.192}$$

In the following, the RIN per unit bandwidth is expressed as the RIN for simplicity. On the derivation of the RIN, see Appendix G.

Figure 5.49 shows the RIN for single-mode operations as a function of the frequency with the light output P as a parameter for three different values of the self-saturation coefficient β. Here, the carrier fluctuations are neglected, that is, $\delta n(t) = \delta\tilde{n}(\omega) = 0$. With an increase in β, the RINs in a frequency range less than 10^8 Hz are drastically reduced. Also, with an increase in P, a frequency region with a flat RIN is expanded, because the self-saturation given by $-\beta P$ cancels out a fluctuation in P.

Figure 5.50 shows the RINs for two-mode operations as a function of the frequency for three values of the light output P_1 of a main mode when the carrier fluctuations are neglected. A parameter is the light extinction ratio P_2/P_1 where P_2 is the light output of a submode. With an increase in P_1, the RINs decrease similarly to single-mode operations. According to P_2/P_1, the RINs show complicated behaviors, because the cross saturation by P_2 also affects a fluctuation in P_1. In a high frequency range over 10^{10} Hz, however, all lines overlap each other in contrast to Fig. 5.49.

Comparing Fig. 5.49 (c) with Fig. 5.50 where the self-saturation coefficient β is common, it can be said that single-mode operations have lower RINs than two-mode operations.

It should be noted that a peak in the RIN, which was already observed experimentally, does not exist when the carrier fluctuations are neglected. The effect of the carrier fluctuations on the RIN will be shown in the next section.

5.13.4 RIN with Carrier Fluctuations

Figure 5.51 shows the RINs for single-mode operations as a function of the frequency with the light output P as a parameter for three values of β when the carrier fluctuations are included. It is found that the RIN has a peak, which has been experimentally observed, as opposed to Fig. 5.49. Thus, the carrier fluctuations are essential for a peak of the RIN. This resonance peak becomes dull when β increases. Therefore, by measuring resonance bandwidths of the RINs, we can determine a value of the self-saturation coefficient β. Comparing Fig. 5.51 with Fig. 5.49, it is also revealed that the RINs are reduced by the carrier fluctuations, and almost-constant RINs are obtained below the resonance frequency.

A resonance frequency f_r in the RIN for single-mode operations is given by

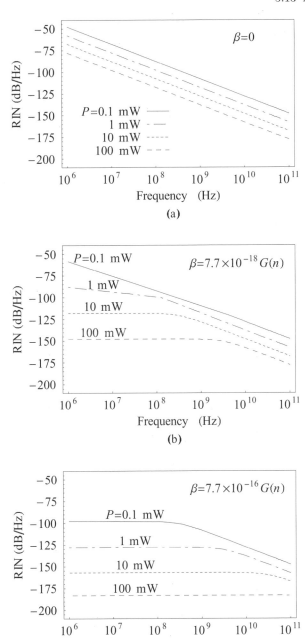

Fig. 5.49. RIN per unit bandwidth for single-mode operations when the carrier fluctuations are neglected

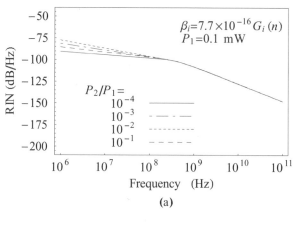

(a)

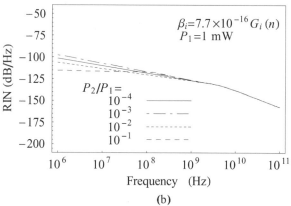

(b)

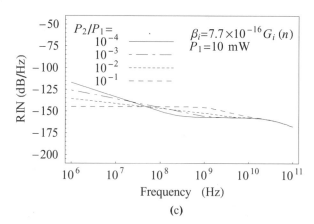

(c)

Fig. 5.50. RIN per unit bandwidth for two-mode operations when the carrier fluctuations are neglected

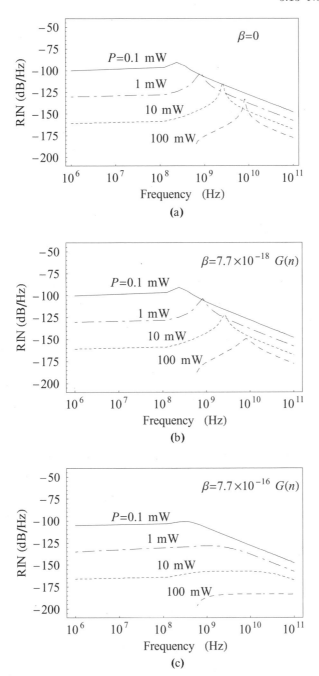

Fig. 5.51. RIN per unit bandwidth for single-mode operations when the carrier fluctuations are included

$$f_{\mathrm{r}} = \frac{1}{2\pi} \sqrt{\left[\frac{\partial G_1}{\partial n}\right]_{n=n_{\mathrm{th}}} \frac{S_{10}}{\tau_{\mathrm{ph1}}} + \beta S_{10} \left(\left[\frac{\partial G_1}{\partial n}\right]_{n=n_{\mathrm{th}}} S_{10} + \frac{1}{\tau_n}\right)} . (5.193)$$

When the light output is low enough or the self-saturation is negligibly small, (5.193) reduces to

$$f_{\mathrm{r}} = \frac{1}{2\pi} \sqrt{\left[\frac{\partial G_1}{\partial n}\right]_{n=n_{\mathrm{th}}} \frac{S_{10}}{\tau_{\mathrm{ph1}}}}, \qquad (5.194)$$

which is the same as the resonance frequency of single-mode semiconductor lasers.

Figure 5.52 shows the RINs for two-mode operations as a function of the frequency for three values of the light output P_1 of a main mode when the carrier fluctuations are considered. A parameter is the light extinction ratio P_2/P_1, where P_2 is the light output of a submode. It is shown that a peak in the RIN is fairly vague in two-mode operations, compared with in single-mode operations. With an increase in P_1, the RINs decrease and show complicated behaviors according to P_2/P_1 due to cross saturation.

Comparing Fig. 5.51 (c) with Fig. 5.52, where the self-saturation coefficient β is common, it is concluded that single-mode operations are superior to simultaneous two-mode operations from the viewpoint of the RINs.

5.13.5 Noises on Longitudinal Modes

(a) Mode Partition Noise

The mode partition noise is observed when a longitudinal mode is selected during multimode operation. This noise is large in low frequencies, as explained in the RINs for two-mode operations. In multimode LDs, such as Fabry-Perot LDs under pulsed operations or the gain guiding LDs, a noise for a total light intensity is comparable to the noise in single-mode LDs. However, a noise for each longitudinal mode in the multimode LDs is much larger than the noise in the single-mode LDs. As a result, the mode partition noise causes a serious problem in mode selective systems such as the optical fiber communication systems.

The cause of the mode partition noise is that the optical gain is randomly delivered to each mode during multimode operation. Therefore, to prevent the mode partition noise, we need single-mode LDs.

(b) Mode Hopping Noise

The mode hopping noise is generated when a longitudinal mode in the single-mode LDs jumps to other modes. This mode hopping is closely related to driving conditions such as temperature and the injection current. At the

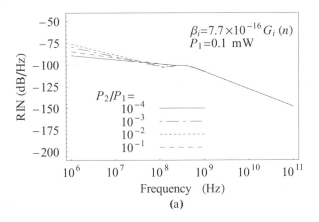

(a)

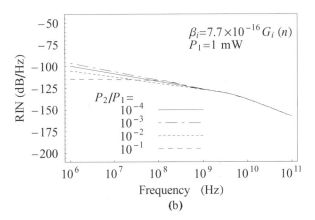

(b)

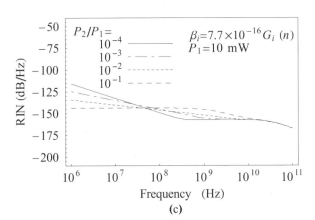

(c)

Fig. 5.52. RIN per unit bandwidth for two-mode operations when the carrier fluctuations are included

time of the mode hopping, random oscillations between multiple modes are repeated and the noise increases due to a difference in light intensities between the relevant modes. When there are two competing modes, the noise is large in a low frequency range below 50 MHz; when there are three or more competing modes, the noise is large up to higher frequencies. Note that the mode partition noise is also large during mode hopping.

For analog systems, such as video discs, the RIN has to be lower than $-140\,\mathrm{dB/Hz}$, and for digital systems, such as compact discs (CD-ROMs), the RIN should be lower than $-120\,\mathrm{dB/Hz}$.

The causes of the mode hopping noise are fluctuations of the spontaneous emission and a propensity for the optical gain to concentrate on the oscillation mode. To avoid the mode hopping noise, two opposite ways, such as single-mode operations and multimode operations are used.

To achieve highly stable single-mode operations, *bistable LDs* or dynamic single-mode LDs are adopted. Bistable LDs containing *saturable absorbers* have hysteresis in *I-L* curves, which suppresses the mode competition. However, it is difficult for them to keep stable single-mode operations with large extinction ratio during modulation. Hence, the bistable LDs are not categorized in the dynamic single-mode LDs.

For systems without mode selectivity, such as video and compact discs, multimode operations are also used to reduce mode hopping noise. Multimode operations have a higher noise level than single-mode operations, but the noise level is stable with changes in temperature or the injection current. Therefore, the maximum noise level is lower than the mode hopping noise. To obtain multimode operations, *high frequency modulations* and *self-pulsations* are used. In high frequency modulation, electric current pulses with a frequency over 600 MHz are injected into the Fabry-Perot LDs and the minimum current is set below the threshold current. The self-pulsations, which are the pulsed operations under DC bias, are obtained in the Fabry-Perot LDs with saturable absorbers or combined structures of the index and gain guidings. Due to the self-pulsations, multimode operations with low coherence take place, which results in stable operations against the optical feedback noise. As a result, they are widely used as light sources for video discs and compact discs.

As described earlier, it is interesting that two contradictory methods, such as single-mode operation and multimode operation are used to suppress the mode hopping noise. In optical fiber communication systems, we need dynamic single-mode LDs, because the optical fibers have mode selectivity due to dispersions. In contrast, mode selectivity does not exist in video and compact discs. In these applications, we have to set up optical systems in a small volume, and we need to reduce optical feedback noise without using *optical isolators*. Therefore, multimode operations are suitable for video and compact discs.

5.13.6 Optical Feedback Noise

Optical feedback noise [42] is generated when a laser light emitted from a
semiconductor laser is fed back to the semiconductor laser itself. A facet of
semiconductor lasers forms *external cavities* with reflective external objects,
such as optical components, optical fibers, and optical discs. These external
cavities and the *internal cavity* of the semiconductor laser compose *coupled
cavities*, which induce optical feedback noise.

The optical feedback noise is noticeable even when the relative feedback
light intensity is on the order of 10^{-6}. Due to optical feedback noise, the light
output characteristics of the semiconductor laser change intricately, according
to the distance between reflective external objects and the semiconductor
laser, the feedback light intensity, and driving conditions. In static or time-
averaged characteristics, the light intensity, the number of lasing modes, and
the light output spectra are modified. In dynamic characteristics, a noise level
and the shape of the light pulse are altered.

(a) Fundamental Equations

We analyze the effect of the feedback lights on the characteristics of semiconductor lasers. Figure 5.53 shows (a) the coupled cavity and (b) its equivalent
model. In the equivalent model, the effect of the external cavity is expressed
by the equivalent amplitude reflectivity r_{eff}.

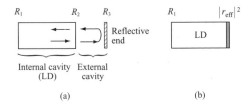

Fig. 5.53. (a) Coupled cavity and (b) its equivalent model

First, we calculate r_{eff}. We assume that the power reflectivities for the
facets of the semiconductor laser are R_1 and R_2, and the power reflectivity
for the external reflector is R_3. Let an angular frequency of a laser light be
Ω and the roundtrip time in the external cavity be τ; then the electric field
of the reflected light $E_r\,e^{i\Omega t}$ is expressed as

$$
E_r\,e^{i\Omega t} = E_i\,e^{i\Omega t}\left[\sqrt{R_2} + (1 - R_2)\sqrt{R_3}\,e^{-i\Omega\tau}\right.
$$
$$
\left. + (1 - R_2)\sqrt{R_2}R_3\,e^{-i2\Omega\tau} + \cdots\right], \qquad (5.195)
$$

where $E_i\,e^{i\Omega t}$ is the electric field of the incident light.

The third- and higher-order terms in $[\cdots]$ of (5.195) correspond to multireflections in the external cavity. Usually, the power reflectivity of the external reflector R_3 is less than several percent. Therefore, we neglect multireflections in the external cavity. As a result, r_{eff} is obtained as

$$r_{\text{eff}} = \frac{E_{\text{r}}}{E_i} = \sqrt{R_2}\left(1 + a\,e^{-i\Omega\tau}\right),$$
$$a = (1 - R_2)\sqrt{\frac{R_3}{R_2}} \ll 1. \tag{5.196}$$

For $R_2 = 32\%$, $R_3 = 1\%$, we have $a = 0.12$.

Secondly, we consider the *decay rate* for the electric field. Let the power decay rate in the semiconductor laser itself be γ_0; then the decay rate for the electric field $\gamma_0/2$ is given by

$$\frac{1}{2}\gamma_0 = \frac{1}{2}\frac{c}{n_{\text{r0}}}\left(\alpha_i + \frac{1}{2L}\ln\frac{1}{R_1 R_2}\right), \tag{5.197}$$

where c is the speed of light in a vacuum, n_{r0} is the effective refractive index of the semiconductor laser in a steady state, α_i is the internal loss in the semiconductor laser, and L is the length of the internal cavity.

Using (5.197), we can write the decay rate for the coupled cavity γ as

$$\frac{1}{2}\gamma = \frac{1}{2}\frac{c}{n_{\text{r0}}}\left(\alpha_i + \frac{1}{2L}\ln\frac{1}{R_1 r_{\text{eff}}^2}\right)$$
$$= \frac{1}{2}\gamma_0 - \kappa\,e^{-i\Omega\tau}, \tag{5.198}$$
$$\kappa = \frac{c}{2n_{\text{r0}}L}a = \frac{c}{2n_{\text{r0}}L}(1 - R_2)\sqrt{\frac{R_3}{R_2}}.$$

Here, κ is the *coupling rate of a feedback light to the semiconductor laser*, and for $R_2 = 32\%$, $R_3 = 1\%$, $n_{\text{r0}} = 3.5$, and $L = 300\,\mu\text{m}$, we have $\kappa = 1.7 \times 10^{10}\,\text{s}^{-1}$. This value is between the decay rate for the carrier concentration $1/\tau_n \sim 10^9\,\text{s}^{-1}$ and that for the photon density $1/\tau_{\text{ph}} \sim 10^{12}\,\text{s}^{-1}$. Using the decay rate γ, we can express an equation for the electric field E as

$$\frac{\text{d}}{\text{d}t}E\,e^{i\Omega t} = \left\{i\omega_N(n) + \frac{1}{2}[G(n) - \gamma]\right\}E\,e^{i\Omega t}, \tag{5.199}$$

where $\omega_N(n)$ is an angular frequency for the Nth-order resonance mode when the carrier concentration is n. Substituting (5.198) into (5.199) results in

$$\frac{\text{d}}{\text{d}t}E(t) = \left\{i\left[\omega_N(n) - \Omega\right] + \frac{1}{2}[G(n) - \gamma_0]\right\}E(t) + \kappa E(t - \tau)\,e^{-i\Omega\tau}, \tag{5.200}$$

where the second term on the right-hand side shows a contribution of the feedback light. Note that this equation includes the phase of the laser light.

Because the optical feedback noise is highly dependent on the phase of the feedback light, we use (5.200) instead of (5.21) to analyze the optical feedback noise.

A rate equation for the carrier concentration n is given by

$$\frac{\mathrm{d}}{\mathrm{d}t}n = \frac{J}{ed} - G(n)|E|^2 - \frac{n}{\tau_n}. \tag{5.201}$$

(b) Effect of the Feedback Light on Static Characteristics

We assume that the electric field E takes a steady-state value. From a real part in (5.200), the amplification rate at the threshold G_{th} is given by

$$G_{\text{th}} \equiv G(n_{\text{th}}) = \gamma_0 - 2\kappa\cos(\Omega\tau). \tag{5.202}$$

From an imaginary part in (5.200), the oscillation angular frequency Ω at the threshold is obtained as

$$\omega_N(n_{\text{th}}) = \Omega + \kappa\sin(\Omega\tau). \tag{5.203}$$

Assuming that the steady-state values $\omega_N(n_{c0})$, Ω_0, and κ_0 satisfy (5.202) and (5.203), we obtain

$$G(n_{c0}) = \gamma_0 - 2\kappa, \tag{5.204}$$

where we have supposed $n_{\text{th}} = n_{c0}$ and $\omega_N(n_{c0}) = \Omega_0$. Because of the carrier lifetime $\tau_n \sim 10^{-9}$ s and the photon lifetime $\tau_{\text{ph}} \sim 10^{-12}$ s, the carrier concentration n does not always take a steady-state value, even though the electric field E is in a steady state. Hence, if we place

$$n_{\text{th}} = n_{c0} + \Delta n, \tag{5.205}$$

then we can write G_{th} as

$$G_{\text{th}} = G(n_{c0}) + \frac{\partial G}{\partial n}\Delta n. \tag{5.206}$$

Substituting (5.206) into (5.202), and using (5.204) result in

$$\Delta n = 2\kappa\left(\frac{\partial G}{\partial n}\right)^{-1}[1 - \cos(\Omega\tau)], \tag{5.207}$$

where we have assumed $\kappa = \kappa_0$ because $1/\tau_n \sim 10^9$ s^{-1} and $\kappa \sim 10^{10}$ s^{-1}.

The resonance angular frequency at the threshold $\omega_N(n_{\text{th}})$ is expressed as

$$\omega_N(n_{\text{th}}) = \omega_N(n_{c0}) - \frac{\omega_N(n_{c0})}{n_{r0}}\frac{\partial n_r}{\partial n}\Delta n. \tag{5.208}$$

Inserting (5.179) and (5.207) into (5.208) leads to

$$\omega_N(n_{\text{th}}) = \omega_N(n_{\text{c}0}) + \alpha\kappa[1 - \cos(\Omega\tau)]. \qquad (5.209)$$

From (5.209), it is found that the spectral linewidth enhancement factor α plays an important role in the optical feedback noise, in addition to the spectral linewidth of the semiconductor lasers. Substituting (5.209) into (5.203), we obtain

$$\omega_N(n_{\text{c}0}) = \Omega + \kappa\sin(\Omega\tau) - \alpha\kappa[1 - \cos(\Omega\tau)]. \qquad (5.210)$$

These results are summarized as follows: When the electric field E takes a steady-state value, the amplification rate at the threshold G_{th} is given by

$$G_{\text{th}} \equiv G(n_{\text{th}}) = \gamma_0 - 2\kappa\cos(\Omega\tau), \qquad (5.211)$$

and the oscillation angular frequency Ω is obtained as

$$\omega_N(n_{\text{c}0}) = \Omega + \kappa\sin(\Omega\tau) - \alpha\kappa[1 - \cos(\Omega\tau)]. \qquad (5.212)$$

The optical feedback noise in semiconductor lasers is larger than that in other lasers because semiconductor lasers have a larger α, a lower R_2, and a shorter L than other lasers, which leads to a larger κ. For example, a gas laser with $\alpha \ll 1$, $R_2 = 98\%$, and $L \sim 1\,\text{m}$ has $\kappa \sim 10^5\,\text{s}^{-1}$, while a semiconductor laser with $\alpha = 1\text{-}7$, $R_2 = 32\%$, and $L \sim 300\,\mu\text{m}$ has $\kappa \sim 10^{10}\,\text{s}^{-1}$.

The terms including trigonometric functions in (5.211) and (5.212) indicate interference between the light in the semiconductor laser and the feedback light. Due to this interference, *hysteresis* accompanies the changes in both the oscillation angular frequency and the light intensity.

Figure 5.54 shows the resonance angular frequency $\omega_N(n_{\text{c}0})$ as a function of the oscillation angular frequency Ω for $\kappa\tau = 1$ and $\alpha = 3$. Here, the arrows indicate the points where Ω jumps due to the nonlinear relationship between the interference condition and the oscillation angular frequency.

Figure 5.55 shows calculated *I-L* curves in DC operations with the spectral linewidth enhancement factor α as a parameter. Here, Joule heating in the active layer is also considered. Hysteresis in the light intensity, which is shown in Fig. 5.55, is experimentally observed.

(c) Effect of Feedback Light on Dynamic Characteristics

Using (5.200) and (5.201), we can analyze the effect of the feedback light on the dynamic characteristics of semiconductor lasers. According to the phase of the feedback light, the dynamic characteristics show complicated behaviors, such as chaos and enhancement or suppression of the relaxation oscillation.

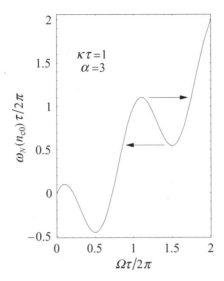

Fig. 5.54. Resonance angular frequency $\omega_N(n_{c0})$ versus oscillation angular frequency Ω

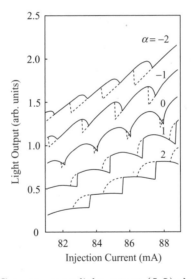

Fig. 5.55. Current versus light output (I-L) characteristics

(d) Enhancement of Noise due to Feedback Light

The most serious problem caused by feedback light is enhancement of noises. Due to the feedback light, the quantum noises increase in a certain frequency region. Moreover, laser oscillations become unstable, and the noise increases in a low frequency range, which is less than several hundred megahertz. The increase in this low frequency noise is caused by the random mode hopping between the longitudinal modes in the internal cavity and those in the external cavity.

(e) Reducing Optical Feedback Noise

To stabilize a longitudinal mode in the internal cavity, we need single-mode LDs such as the DFB-LDs and the bistable LDs. To suppress the interference between the feedback light and the internal light, we should reduce coherence of the laser light by high frequency modulation or self-pulsation. To decrease the feedback light intensity, we require a low coupling rate κ, which is achieved by large facet reflectivity and a long cavity as shown in (5.198). However, a large reflectivity leads to a low light output, and a long cavity results in a large threshold current. Therefore, the optical isolators are used to decrease the feedback light intensity in the optical fiber communication systems, but the cost and size of the optical systems increase.

5.14 Degradations and Lifetime

Degradation of semiconductor lasers means that the threshold currents increase and the external quantum efficiencies decrease, as shown in Figure 5.56. If the degradation continues, CW laser operations stop.

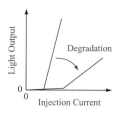

Fig. 5.56. Degradation of semiconductor lasers

The causes of the degradation, which depend on the properties of the materials, are propagation of the *crystal defects*, changes in the surface condition, destruction of facets, generation of the point defects, degradation of the *ohmic contacts* or the contact layers with the *heat sinks*, and so on. The speed of degradation is enhanced with increases in the temperature, the light output, and the injection current.

In *automatic power control* (APC), the *lifetime* of semiconductor lasers is usually defined as a time when the operating current increases up to twice the initial value. In InGaAsP/InP LDs used in trunk-line optical fiber communication systems, the expected lifetime is more than 27 years.

5.14.1 Classification of Degradations

(a) Catastrophic Optical Damage

When the temperature at a facet of a semiconductor laser increases up to a melting point, the light output from the melted facet rapidly falls, which is referred to as the *catastrophic optical damage* (COD). The destructed facet never returns to its original condition. Figure 5.57 schematically shows the catastrophic optical damage. In AlGaAs/GaAs LDs, the critical light output density of the COD in CW operations is on the order of 10^6 W/cm^2. If the active layer is 0.2 µm thick and 5 µm wide, the critical light output of the COD is only 10 mW.

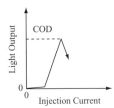

Fig. 5.57. Catastrophic optical damage

The process of the catastrophic optical damage is as follows: Crystal surfaces have a lot of surface energy levels, which lead to nonradiative recombinations. As a result, the carrier concentration in the vicinity of the facet is too low to generate optical gain, and this region absorbs laser light. Due to absorption of the laser light, the temperature at the facet increases and the bandgap shrinks. Hence, the absorption coefficient at the facet further increases, which accelerates absorption of the laser light and Joule heating of the facet. Such positive feedback increases the temperature at the facet up to the melting point (about 1500°C in AlGaAs), and the facet melts.

To suppress the COD, the vicinity of the facet should be transparent for the laser light. Figure 5.58 shows *window structures*, which use a property that a bandgap in n-AlGaAs is larger than that in p-AlGaAs. In CW operation, the critical light outputs of the window structure LDs are several times higher than those of LDs without windows. In pulsed operations, the COD level is improved by a factor of ten.

In materials with a low nonradiative recombination rate at the surface or high thermal conductivity, the COD levels are high. For example, in InGaAsP/InP LDs, the COD is not observed even in large light output over 100 mW.

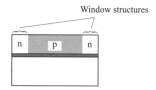

Fig. 5.58. Window-structure LD

(b) Dark Line Defects

The *dislocations* cause *dark line defects*, which are named after the dark lines observed in light emission patterns. The dark line defects propagate by growth of the dislocations with repetitive emissions and absorptions of the point defects or extension of the dislocations due to movement of the slipped dislocations. Once the dark lines are generated, the nonradiative recombinations and the absorptions increase.

Among the dislocations, which are the sources of the dark lines, there exist (1) the penetration dislocations, that is, the dislocations in a substrate extended to the epitaxial layers; (2) the dislocations generated at the interfaces of the heterojunctions; (3) the dislocations based on some deposits, which were formed during epitaxial growth; (4) the misfit dislocations generated above critical thickness due to lattice mismatching; and (5) the dislocations induced by external stresses. To suppress these dislocations, we need to use high-quality substrates with few dislocations and optimize epitaxial growth conditions. It should be noted that in InGaAsP/InP LDs, the dislocations grow slowly and the dark lines do not appear frequently.

(c) Facet Degradation

In AlGaAs/GaAs LDs, even when they are placed in N_2-ambient or air-proof packages, the facets are easily oxidized by a fraction of remaining oxygen. In contrast, the facets of InGaAsP/InP LDs are not readily oxidized. Figure 5.59 shows oxidization of the facets. If the facet is oxidized as shown in Fig. 5.59 (a), the laser light is scattered, which leads to a large radiation angle. Also, the oxidized region and its neighbor have many point defects and impurities, such as oxygen, which become nonradiative recombination centers. Hence, the threshold current increases, and the external quantum efficiency decreases. To avoid oxidization of the facets, dielectric thin films, such as SiO_2, Al_2O_3, and Si_3N_4, are often coated on the facet, as illustrated in Fig. 5.59 (b). It should be noted that these dielectric thin films cannot reduce the dark lines, but the window structures can suppress oxidization of the facets.

Fig. 5.59. Oxidation of a facet: (a) degradation and (b) protection

5.14.2 Lifetime

There are two definitions of the lifetime according to the driving conditions. In *automatic power control* (APC), the lifetime is a time in which a predetermined light output is not obtained. Considering the degradations in the longitudinal modes, transverse modes, and relaxation oscillation, the lifetime is often defined as a time when the operating current increases up to the double of the initial value. In *automatic current control* (ACC), the lifetime is frequently defined as a time when the light output decreases down to half of the initial value.

In manufacturing processes of semiconductor lasers, *intentionally accelerating degradation tests* called the *screening tests* are widely used. It is experimentally confirmed that after the screening tests low degradation LDs have long lifetimes.

6 Dynamic Single-Mode LDs

6.1 Introduction

In order to transmit signals, we modulate a laser light by direct modulation or external modulation. Note that direct modulation is superior to external modulation in terms of cost.

Dynamic single-mode LDs are semiconductor lasers, which show single-longitudinal-mode operations in direct modulation. Because the optical fibers have dispersions, dynamic single-mode LDs are key devices for long-haul, large-capacity optical fiber communication systems. Therefore, the DFB-LDs, the DBR-LDs, the surface emitting LDs, and the cleaved coupled cavity (C^3) LDs have been developed. Because the optical gain spectra of semiconductor lasers have a linewidth on the order of 10 nm, the optical cavities play important roles to select only one resonance mode for dynamic single-mode operations. It should be noted that the bistable LDs with saturable absorbers are not categorized into dynamic single-mode LDs, because they show single-mode operations only in a steady state and lead to multimode operation in direct modulation.

6.2 DFB-LDs and DBR-LDs

Distributed feedback (DFB) LDs and distributed Bragg reflector (DBR) LDs have diffraction gratings in the optical waveguides to achieve high mode selectivity. As shown in Fig. 6.1, the DFB-LDs have the active layers in the corrugated regions, and the DBR-LDs do not have the active layers in the corrugated regions.

The DFB-LDs and DBR-LDs periodically modulate their complex refractive indexes to achieve optical feedback. The reflection points of the Fabry-Perot LDs are only the facets, while those of the DFB-LDs and the DBR-LDs are distributed all over the corrugated regions.

Fig. 6.1. (a) DFB-LD and (b) DBR-LD

6.2.1 DFB-LDs

(a) Index-Coupled DFB-LDs

In an *index-coupled grating*, only a real part of the complex refractive index is periodically modulated and an imaginary part is uniform. The real part $n_{\mathrm{r}}(z)$ is expressed as

$$n_{\mathrm{r}}(z) = n_{\mathrm{r0}} + n_{\mathrm{r1}} \cos(2\beta_0 z + \Omega). \tag{6.1}$$

Here, z is a position, Ω is the grating phase at $z = 0$, and β_0 is written as

$$\beta_0 = \frac{\pi}{\Lambda}, \tag{6.2}$$

where Λ is the grating pitch.

Figure 6.2 shows the index-coupled grating. If we form corrugations on the interfaces of two layers with refractive indexes n_{A} and $n_{\mathrm{B}}(\neq n_{\mathrm{A}})$, we can periodically modulate the refractive index. Due to this periodical modulation of the refractive index, a forward running wave and a backward running wave are coupled to each other, which is indicated by the grating coupling coefficient κ given by

$$\kappa = \frac{\pi n_{\mathrm{r1}}}{\lambda_0}. \tag{6.3}$$

Here, we put $\alpha_1 = 0$ in (4.34). The grating coupling coefficient κ is an important parameter, which represents resonance characteristics of the DFB-LDs and the DBR-LDs.

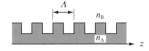

Fig. 6.2. Index-coupled grating

(i) Uniform Grating

We consider the oscillation condition of the DFB-LD with a uniform grating, in which both the depth and the pitch are constant all over the corrugated region. As shown in Fig. 6.3, we assume that both facets are antireflection (AR) coated, and reflections at both facets are negligibly low.

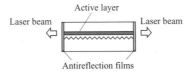

Fig. 6.3. DFB-LD with a uniform grating

When the transmissivity in (4.48) is infinity, the DFB-LD starts to lase. As a result, from (4.45) and (4.47), the threshold condition of the DFB-LD with a uniform grating is obtained as

$$\cosh(\gamma L) - \frac{\alpha_0 - i\delta}{\gamma} \sinh(\gamma L) = 0, \tag{6.4}$$

where α_0 is the amplitude gain coefficient, L is the cavity length, and

$$\gamma^2 = (\alpha_0 - i\delta)^2 + \kappa^2, \tag{6.5}$$

$$\delta = \frac{\beta^2 - \beta_0{}^2}{2\beta_0} \simeq \beta - \beta_0. \tag{6.6}$$

Here, β is the propagation constant of a light.

Figure 6.4 shows transmission spectra of a uniform grating for four values of the optical gain, where α_{th} is the threshold gain. As found in Fig. 6.4, the DFB-LD with a uniform grating oscillates in two modes when the reflectivities of both facets are negligibly low.

When the reflectivities of both facets are not negligible as in the cleaved facets, the oscillation modes show complicated behaviors according to the grating phases Ω at the facets. For a certain grating phase, single-mode operations are obtained. However, the grating pitch is as short as about 0.2 μm, and it is almost impossible to control the cleaved position of the gratings in manufacturing. Therefore, it is difficult to obtain single-mode operations in the DFB-LDs with a uniform grating, and a yield for single-mode operations is less than several percent.

To achieve stable single-mode operations with a high yield, the phase-shifted DFB-LDs and the gain-coupled DFB-LDs have been developed.

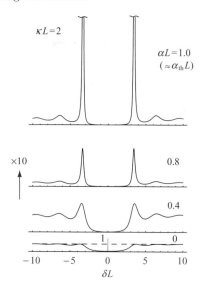

Fig. 6.4. Transmission spectrum for a uniform grating

(ii) Phase-Shifted Grating

As shown in Fig. 6.5, the corrugations of the phase-shifted grating are shifted, whereas the corrugations of the uniform gratings continue as indicated by a broken line. Here, the phase shift is expressed as $-\Delta\Omega$ according to the definition of the refractive index in (6.1).

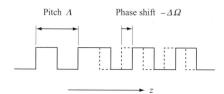

Fig. 6.5. Phase-shifted grating

Let us consider an analytical model that consists of two regions and includes the phase shift as a phase jump at the interface, as shown in Fig. 6.6. We express a transfer matrix for region 1 with length of L_1 as $\boldsymbol{F}_1$, and that for region 2 with length of L_2 as $\boldsymbol{F}_2$. Here, we assume that both the pitch and the depth of the corrugations are constant except in the phase-shifted region, and the optical gain is uniform all over the grating.

We suppose that both facet reflectivities are negligibly low. If we write the grating phase Ω at the left edge of region 1 as θ_1, the grating phase θ_2 at the right edge of region 1 is given by

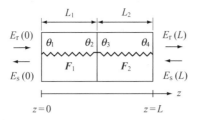

Fig. 6.6. Analytical model for the phase-shifted grating

$$\theta_2 = \theta_1 + 2\beta_0 L_1. \tag{6.7}$$

Because of the phase shift $\Delta\Omega$ at the interface of regions 1 and 2, the grating phase θ_3 at the left edge of region 2 is expressed as

$$\theta_3 = \theta_2 + \Delta\Omega = \theta_1 + 2\beta_0 L_1 + \Delta\Omega. \tag{6.8}$$

The threshold condition is given by $F_{11} = 0$, where F_{11} is a matrix element of $\boldsymbol{F} = \boldsymbol{F}_1 \times \boldsymbol{F}_2$ and is written as

$$\left[\cosh(\gamma L_1) - \frac{\alpha_0 - \mathrm{i}\,\delta}{\gamma}\sinh(\gamma L_1)\right]\left[\cosh(\gamma L_2) - \frac{\alpha_0 - \mathrm{i}\,\delta}{\gamma}\sinh(\gamma L_2)\right]$$
$$+ \frac{\kappa^2}{\gamma^2}\sinh(\gamma L_1)\sinh(\gamma L_2)\,\mathrm{e}^{\mathrm{i}\Delta\Omega} = 0. \tag{6.9}$$

When the phase-shifted position is located at a center of the optical cavity ($L_1 = L_2 = L/2$), (6.9) reduces to

$$\left[\cosh(\gamma L) - \frac{\alpha_0 - \mathrm{i}\,\delta}{\gamma}\sinh(\gamma L)\right]$$
$$+ \frac{\kappa^2}{\gamma^2}(\mathrm{e}^{\mathrm{i}\Delta\Omega} - 1)\left[\sinh\left(\frac{\gamma L}{2}\right)\right]^2 = 0. \tag{6.10}$$

The first term in (6.10) is the same as the left-hand side in (6.4) for the DFB-LD with a uniform grating. The second term in (6.10) represents an effect of the phase shift. Figure 6.7 shows calculated results of the amplitude threshold gain $\alpha_{\mathrm{th}} L$ as a function of $\delta L = \delta \times L$ for three phase-shift values.

Laser oscillation starts at a mode with the lowest $\alpha_{\mathrm{th}} L$. If a difference in the threshold gain of the oscillation mode and that of other modes is large enough, highly stable single-mode operations are obtained. For the phase shift $\Delta\Omega = 0$ as in the uniform grating, two modes have a common lowest threshold gain $\alpha_{\mathrm{th}} L$, as indicated by open circles. Therefore, $\Delta\Omega = 0$ results in two mode operations. For the phase shift $-\Delta\Omega = \pi/2$ shown by open triangles, there is only one mode whose threshold gain is the lowest, and a single-mode operation is expected. For the phase shift $-\Delta\Omega = \pi$ represented by closed circles, there exists only one mode whose threshold gain is the

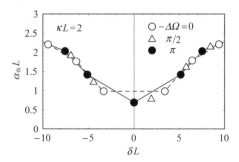

Fig. 6.7. Calculated amplitude threshold gain $\alpha_{\mathrm{th}}L$ as a function of $\delta L = \delta \times L$ for three phase-shift values

lowest at Bragg wavelength. Note that the difference in the lowest threshold gain and the second-lowest one is largest for $-\Delta\Omega = \pi$, which leads to the most stable single-mode operations.

From (4.41), Bragg wavelength λ_B in a vacuum, which satisfies $\delta = 0$, is given by

$$\lambda_B = \frac{2n_{r0}\Lambda}{m}, \tag{6.11}$$

where Λ is the grating pitch and m is a positive integer, which is called *order of diffraction*.

When the phase shift $-\Delta\Omega$ is π in the first-order gratings ($m = 1$), the corrugations are shifted by $\Lambda/2$. From (6.11), we have

$$\frac{\Lambda}{2} = \frac{\lambda_B}{4n_{r0}} = \frac{\lambda_m}{4}, \quad \lambda_m = \frac{\lambda_B}{n_{r0}}, \tag{6.12}$$

where λ_m is a wavelength in a material. From (6.12), the phase shift of π corresponds to a quarter of a wavelength in a material. Therefore, the phase-shifted grating with $-\Delta\Omega = \pi$ is often called a **$\lambda/4$-*shifted grating*** or a *quarter-wavelength-shifted grating*.

Figure 6.8 shows transmission spectra of the $\lambda/4$-shifted grating for four values of the optical gain. As found in Fig. 6.8, the $\lambda/4$-shifted DFB-LD oscillates at Bragg wavelength located at the center of the stop band and shows highly stable single-mode operations when the reflectivities at both facets are negligibly low.

According to the light intensity distribution along the cavity axis, the radiative recombinations are altered and spatial distribution of the carrier concentration is modified. This phenomenon is designated as the *spatial hole burning*, and the distribution of the refractive index changes with that of the carrier concentration, which may lead to a change in the phase shift. Therefore, a grating with a phase shift slightly altered from π is used to achieve

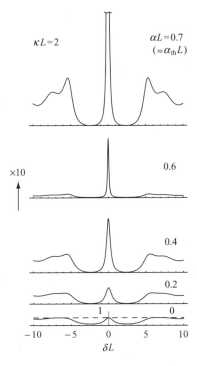

Fig. 6.8. Transmission spectrum for a $\lambda/4$-shifted grating

$-\Delta\Omega = \pi$ at the operating condition, or the chirped grating is adopted to reduce the spatial hole-burning.

When there are reflections at both facets, as in the cleaved facets, the oscillation modes of the phase-shifted DFB-LDs show complicated behaviors according to the grating phases at the facets, which is similar to the DFB-LD with a uniform grating. In addition, stability of single-mode operations is lower than that of the phase-shifted DFB-LD without reflections at the facets. Therefore, to achieve highly stable single-mode operations in the phase-shifted DFB-LD, we need antireflection coated facets or window structures to reduce reflections at the facets.

Due to the high stability of single-mode operations, the index-coupled DFB-LDs are used in long-haul, large-capacity optical fiber communication systems such as trunk-line optical fiber cable systems in Japan and submarine optical fiber cable systems between Japan and the United States. Because the phase-shifted DFB-LD has been developed after the DFB-LD with a uniform grating, both DFB-LDs have been practically used. However, it is needless to say that stability and reproducibility of single-mode operations in the phase-shifted DFB-LD are better than those in the DFB-LD with a uniform grating.

Finally, we briefly explain the polarizations of the laser lights in the DFB-LDs. Because the grating coupling coefficient κ_{TE} for the TE mode is larger than that for the TM mode κ_{TM}, the threshold gain for the TE mode is lower than that for the TM mode. Therefore, the DFB-LDs start to lase in the TE mode. However, a difference in the threshold gains between the TE and TM modes is smaller than that of the Fabry-Perot LDs if the reflectivities at both facets are negligibly low. Hence, to stabilize the polarization of a laser light, quantum well active layers are frequently adopted to obtain polarization-dependent optical gains, which will be described in Chapter 7.

(b) Gain-Coupled DFB-LDs

In gain-coupled DFB-LDs [43], the optical gain or optical loss is periodically modulated along the cavity axis. They are characterized by stable single-mode operations even without the phase-shifted gratings and antireflection films coated on the facets. Also, they are less sensitive to feedback lights than the index-coupled DFB-LDs. However, there are still problems in fabrication methods and reliability of the devices.

When only the optical gain or loss is periodically modulated, the grating coupling coefficient κ is given by

$$\kappa = i\,\frac{\alpha_1}{2}, \tag{6.13}$$

where α_1 is a deviation from a steady-state amplitude gain coefficient α_0. If both the real and imaginary parts of the complex refractive index are periodically modulated with the same phase, κ is given by (4.34).

6.2.2 DBR-LDs

In DBR-LDs, the optical gain regions and corrugated regions are separated from each other. The corrugated regions function as reflectors. As a result, a DBR-LD with a DBR and cleaved facets and one with two DBRs have been developed, as shown in Fig. 6.9.

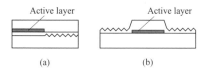

Fig. 6.9. DBR-LDs: (a) DBR and cleaved facets and (b) two DBRs

Figure 6.10 shows the reflectivities of the cleaved facet and the DBR. The reflectivity R_0 for the cleaved facet is considered to be almost independent of a light wavelength, although R_0 is slightly modified by the material dispersion. In contrast, the reflectivity R_1 for the DBR is high only within the stop band. Because the resonance condition is not always satisfied at Bragg wavelengths, the DBR-LDs do not always lase at Bragg wavelengths.

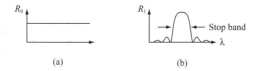

(a)   (b)

Fig. 6.10. Reflectivities of the (a) cleaved facet and (b) DBR

Note that the DFB-LDs are superior to the DBR-LDs in both stability of single-mode operations and light-output intensity.

As wavelength tunable LDs, DFB- or DBR-LDs with phase-control sections have been demonstrated. If these LDs are biased just below the threshold, they function as wavelength tunable resonant optical amplifiers (optical filters).

6.3 Surface Emitting LDs

6.3.1 Vertical Cavity Surface Emitting LDs

Figure 6.11 shows the schematic structure of a *vertical cavity surface emitting LD* (VCSEL) [44], in which the active layers and the cladding layers are sandwiched by the reflectors with multilayers.

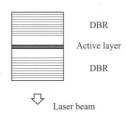

Fig. 6.11. Vertical cavity surface emitting LD

A laser light propagates along a normal to the surface of the epitaxial layers. Therefore, the length of the optical gain region is equal to the active layer thickness, which is on the order of ten nanometers to several micrometers. In contrast, the length of the optical gain region in conventional Fabry-Perot LDs or DFB-LDs is on the order of $300\,\mu$m. As a result, to achieve a low threshold in VCSELs, we need high reflective reflectors, for example, with the power reflectivity of 99.5%. Such a high reflectivity is achieved by multilayer reflectors, which consist of alternately stacked layers with different refractive indexes. As explained in Chapter 4, the operating principles behind the multilayer reflectors and the waveguide DBRs are common, and the multilayer reflectors used in the VCSELs are also called DBRs.

Note that the optical cavities of the VCSELs are short enough to obtain dynamic single-mode operations. For example, when the refractive index is 3.5 and the cavity length is 4 μm, the oscillation wavelength is 1 μm and the free spectral range is 36 nm from (4.15). This value is larger than the linewidth of the optical gain spectra such as 10 nm. As a result, there exists only one longitudinal mode in the gain spectrum.

Compared with other dynamic single-mode LDs, the VCSELs have the following additional features:

1. It is possible to fabricate monolithic optical cavities, for which cleaving is not needed.
2. It is easy to test devices on wafers before pelletizing.
3. It is easy to couple a laser light to optical components, because circular beams with small radiation angles are obtained.
4. It is possible to integrate devices by stacking.
5. Because horizontal sizes can be less than several tens of micrometers,
   a) extremely low threshold currents are expected and
   b) the structures are suitable for high-density two-dimensional arrays.

Based on these features, the VCSELs take a lot of attention as key devices for parallel optical information processing and parallel lightwave transmissions. Also, optical functional devices based on the VCSELs have been demonstrated.

6.3.2 Horizontal Cavity Surface Emitting LDs

In the horizontal cavity surface emitting LDs [45, 46], some structures are added to conventional edge-emitting semiconductor lasers to emit laser lights along a normal to the epitaxial layer planes. Figures 6.12 (a) and (b) show the structures where a part of the epitaxial layers is etched off to form reflection surfaces to emit laser lights. However, fabrication methods to form low-damage reflectors have yet not been established. Figure 6.12 (c) has a second-order grating, and a first-order diffracted laser beam is emitted upward. However, the beam divergence angle is wide, and the diffraction efficiency of the second-order grating is low.

6.4 Coupled Cavity LDs

The coupled cavity LDs use optical feedback to achieve single-mode operations. As shown in Fig. 6.13, the external cavity LD with a mirror and

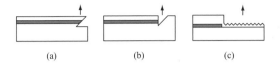

Fig. 6.12. Horizontal cavity surface emitting LDs

with a diffraction grating and the cleaved coupled cavity (C^3) LD have been developed.

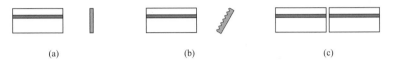

Fig. 6.13. Coupled cavity LDs: (a) external cavity LD (mirror), (b) external cavity LD (grating), and (c) cleaved coupled cavity LD

The external cavity LDs have long cavities, 10 cm or more, and their cavities are sensitive to mechanical vibrations. Also, only a part of the optical cavity contributes to modulation of laser lights, which leads to low modulation efficiency. The C^3 LDs, in which the Fabry-Perot LDs are placed in series, have solved all the problems in the external cavity LDs. However, we need precise control of the injection currents into the two constituent Fabry-Perot LDs to obtain stable single-mode operations, and complicated driving electronic circuits are required.

Among dynamic single-mode LDs, only the DFB-LDs are widely used in practical applications for the reasons described earlier. There are various approaches to solve problems. However, we should keep in mind that practically used technologies work for a reason.

7 Quantum Well LDs

7.1 Introduction

Quantum well (QW) LDs [47,48] are semiconductor lasers whose active layers take quantum well structures. A QW-LD with one potential well is called the *single quantum well* (SQW) LD, and that with plural QWs is named the *multiple quantum well* (MQW) LD.

As explained in Chapter 1, the density of states of the QW is a step function of the energy. Therefore, excellent characteristics, such as a low threshold current, a high differential quantum efficiency, high-speed modulation, low chirping, and a narrow spectral linewidth are simultaneously obtained in the QW-LDs.

7.2 Features of Quantum Well LDs

7.2.1 Configurations of Quantum Wells

Figure 7.1 shows the configurations of various QWs at the band edge. Figure 7.1 (a) illustrates the SQW, in which the optical confinement factor Γ_a is reduced with a decrease in the active layer thickness L_z, as shown in Fig. 5.9. This reduction in Γ_a leads to a drastic increase in the threshold current density J_{th} as illustrated in Fig. 5.10. Therefore, to obtain a larger Γ_a than that of the SQW, structures (b)–(e) have been developed.

Figure 7.1 (b) shows the MQW, in which the optical confinement factor Γ_a is enlarged by forming multiple QWs in the optical confinement region. However, due to the energy barriers between the adjacent QWs, the carrier injection efficiency decreases with an increase in the propagation distance of the carriers. Hence, it is difficult to achieve uniform carrier distribution all over the QWs.

To improve the carrier injection efficiency and the uniformity of the carrier distribution in the MQW, the *modified MQW* in Fig. 7.1 (c) has been developed. The energy barriers between the adjacent QWs are lower than those in the cladding layers, which results in a high carrier injection efficiency and uniform carrier distribution all over the QWs.

The *separate confinement heterostructure* (SCH) exhibited in Fig. 7.1 (d) has been demonstrated to increase the optical confinement factor Γ_a in the SQW. In the materials conventionally used for semiconductor lasers, the refractive index increases with a decrease in the bandgap energy. Using this property, the energy potential is modified by two steps in the SCH structure. The outer potential confines a light to the QW active layer by the refractive index distribution, while the inner potential confines the carriers by the energy barriers. Because the potentials to confine the light and the carriers are separate, this structure is called the SCH.

Figure 7.1 (e) illustrates the *graded index* SCH (GRIN-SCH) whose potential and refractive index distributions of the outer region are parabolic. In the GRIN-SCH, the optical confinement factor Γ_a is proportional to the active layer thickness L_z with a small L_z, while Γ_a in the SQW is in proportion to $L_z{}^2$. Therefore, when the active layer is thin, a relatively large Γ_a is obtained in the GRIN-SCH.

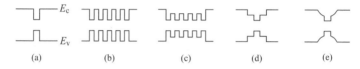

Fig. 7.1. Configurations of quantum wells: (a) SQW, (b) MQW, (c) modified MQW, (d) SCH, and (e) GRIN-SCH

7.2.2 Characteristics of QW-LDs

(a) Low Threshold Current

It is important to achieve a low threshold current I_{th} in semiconductor lasers, because a low I_{th} leads to low power consumption. The density of states $\rho(E)$ per unit energy per unit area in the SQW is written as

$$\rho(E) = \sum_{n=1}^{\infty} \frac{m^*}{\pi\hbar^2} H(E - \epsilon_n), \tag{7.1}$$

where E is the energy of the carrier, m^* is the effective mass of the carrier, $\hbar$ is the Dirac's constant, and $H(x)$ is the *Heaviside function* or a *step function* given by

$$H(x) = \begin{cases} 0 & : & x < 0, \\ 1 & : & x \geq 0. \end{cases} \tag{7.2}$$

If the barrier height or the barrier thickness is too large for each QW to be independent of each other, the energy level of the nth subband ϵ_n is expressed as

$$\epsilon_n = \frac{(n\pi\hbar)^2}{2m^*{L_z}^2}.$$ (7.3)

When the barrier height or barrier thickness is large enough to separate each QW, the density of states $\rho(E)$ per unit energy per unit area in the MQW is obtained as

$$\rho(E) = N \sum_{n=1}^{\infty} \frac{m^*}{\pi\hbar^2} H(E - \epsilon_n),$$ (7.4)

where N is the number of QWs.

When the barrier height or barrier thickness is small, the wave functions for the carriers penetrate into the adjacent QWs. Therefore, the QWs are coupled to each other, and degeneracy is removed to generate N quantum levels per degenerated quantum level. In this case, the density of states $\rho(E)$ is given by

$$\rho(E) = \sum_{n=1}^{\infty} \sum_{k=1}^{N} \frac{m^*}{\pi\hbar^2} H(E - \epsilon_{nk}),$$ (7.5)

where ϵ_{nk} $(k = 1, 2, \cdots, N)$ is the energy of the quantum level, which is produced by the removal of degeneracy.

As described earlier, the densities of states in the QW-LDs are step functions, which results in narrow optical gain spectra. Hence, the optical gain concentrates on a certain energy, and the peak gain is enhanced. As a result, a threshold current density J_{th} as low as 1/3 of that of a bulk double heterostructure (DH) LD was obtained in a QW-LD.

Here, we consider the linear optical gains of the QW-LDs. We assume that the nonradiative recombinations are negligibly low. Under the $\mathbf{k}$-selection rule, the linear optical gain g (in units of cm^{-1}) is written as

$$\frac{c}{n_{\text{r}}} g(E, n) = \frac{\omega}{n_{\text{r}}^2} \chi_{\text{I}}(E, n),$$

$$\frac{\omega}{n_{\text{r}}^2} \chi_{\text{I}}(E, n) = \int \sum_{n=0}^{\infty} \sum_{j=l,h} \rho_{\text{red},n}^{j} [f_{\text{c}}(E_{\text{c},n}) - f_{\text{v}}(E_{\text{v},n}^{j})] \hat{\chi}_{\text{I}}^{n,j}(E, \epsilon) \, \mathrm{d}\epsilon,$$ (7.6)

$$\hat{\chi}_{\text{I}}^{n,j}(E, \epsilon) = \frac{e^2 h}{2m^2 \varepsilon_0 n_{\text{r}}^2 E_{\text{g}}} |M_{n,j}(\epsilon)|_{\text{ave}}^2 \frac{\hbar/\tau_{\text{in}}}{(E - \epsilon)^2 + (\hbar/\tau_{\text{in}})^2},$$

where c is the speed of light in a vacuum; n_{r} is the effective refractive index of the semiconductor laser; E is the photon energy; ω is the angular frequency of a light; χ_{I} is an imaginary part of the *relative electric susceptibility*, and $j = \text{h}, \text{l}$ represents the heavy hole (h) and the light hole (l), respectively. In the second equation of (7.6), $\hat{\chi}_{\text{I}}^{n,j}(E, \epsilon)$ is the imaginary part of the relative electric susceptibility for a photon with the energy E and an electron-hole pair with the energy ϵ. In the third equation of (7.6), e is the elementary charge, h is Planck's constant, m is the electron mass in a vacuum, ε_0 is permittivity in a vacuum, E_g is the bandgap energy, $|M_{n,j}(\epsilon)|^2_{\text{ave}}$ is a square of the momentum matrix element, and τ_{in} is the *intraband relaxation time*. The reduced density of states for the nth subband $\rho^j_{\text{red},n}$ is given by

$$\rho^j_{\text{red},n} = \left(\frac{1}{\rho_{c,n}} + \frac{1}{\rho^j_{v,n}} \right)^{-1}, \tag{7.7}$$

where $\rho_{c,n}$ and $\rho^j_{v,n}$ are the density of states for the nth subband in the conduction band and that in the valence bands, respectively; $f_c(E_{c,n})$ and $f_v(E^j_{v,n})$ are the Fermi-Dirac distribution function of the conduction band and that of the valence bands, respectively. Here, $E_{c,n}$ and $E^j_{v,n}$ are the energy of the conduction band and that of the valence bands, respectively, and are expressed as

$$E_{c,n} = \frac{m_c \epsilon^j_{v,n} + m^j_v E + m^j_v \epsilon_{c,n}}{m_c + m^j_v}, \tag{7.8}$$

$$E^j_{v,n} = \frac{m_c \epsilon^j_{v,n} - m_c E + m^j_v \epsilon_{c,n}}{m_c + m^j_v}, \tag{7.9}$$

where m_c and m^j_v are the effective mass of the electron and that of the holes, respectively; and $\epsilon_{c,n}$ and $\epsilon^j_{v,n}$ are the energy of the nth subband in the conduction band and that in the valence bands, respectively. Figure 7.2 (a) shows the *modal gain* $g_{\text{mod}} = \Gamma_a g$ as a function of the injection current density J with the number of QWs N as a parameter. Here, Γ_a is the optical confinement factor of the active layer.

It is found that g_{mod} takes a different value according to N for a common current density. Hence, the number of QWs to minimize J_{th} depends on the optical loss in the optical cavity, which is equal to the threshold gain. Also, in the SQW ($N = 1$), *gain flattening* is observed with an increase in the injection current. As shown in Fig. 7.2 (b), the cause of the gain flattening is that the density of states is a step function. Here, E_L is a lasing photon energy.

(b) Characteristic Temperature

Because the densities of states in the QWs are step functions, changes in the characteristics of the QW-LDs with the temperature are expected to be small. However, even in a bulk DH-LD, a characteristic temperature T_0 as high as $200\,\text{K}$ is obtained, and an advantage of the QW-LDs over the bulk DH-LDs has not been proved yet.

(c) Anisotropic Optical Gain

A difference in the optical gains for the TE and TM modes is about $20\,\text{cm}^{-1}$ in the bulk DH-LDs, while it is as large as $\sim 140\ \text{cm}^{-1}$ in the QW-LDs. In the

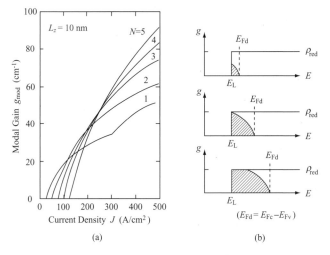

(a)   (b)

Fig. 7.2. Modal gain and gain flattening

bulk DH-LDs, an *anisotropic optical gain* is generated by the configurations of the optical waveguides, because the optical confinement factor Γ_a of the TE mode is larger than that of the TM mode. In the QW-LDs, the anisotropic optical gain is produced by the selection rule of the optical transitions, which will be explained in the following.

As described in Chapter 1, in the bulk structures, the heavy hole (hh) band with $m_j = 3/2$ and the light hole (lh) band with $m_j = 1/2$ are degenerate at $\boldsymbol{k} = 0$, while in the QWs this degeneracy is removed. Figure 7.3 schematically shows the energies of the valence band. Here, E_{hh1} and E_{hh2} (solid lines) represent the heavy hole bands; E_{lh1} and E_{lh2} (broken lines) indicate the light hole bands where a subscript 1 or 2 is a principal quantum number n.

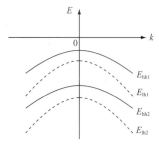

Fig. 7.3. Valence band in a one-dimensional QW

It should be noted that the effective masses of the holes in the QWs are dependent on directions. As shown in Fig. 7.4, if we select a quantization axis as the z-axis, the effective mass along the z-axis of the heavy hole $m_{z,\text{hh}}{}^*$ is larger than that of the light hole $m_{z,\text{lh}}{}^*$, while the effective mass on the xy-plane of the heavy hole $m_{xy,\text{hh}}{}^*$ is smaller than that of the light hole $m_{xy,\text{lh}}{}^*$. This result is summarized as

$$\text{along the } z\text{-axis: } m_{z,\text{hh}}{}^* > m_{z,\text{lh}}{}^*,$$
$$\text{on the } xy\text{-plane: } m_{xy,\text{hh}}{}^* < m_{xy,\text{lh}}{}^*.$$

Because a wave function of the heavy hole distributes in the xy-plane corresponding to the p_x- or p_y-like orbital, the heavy hole moves on the xy-plane more easily than along the z-axis, which leads to $m_{xy,\text{hh}}{}^* < m_{z,\text{hh}}{}^*$. In contrast, a wave function of the light hole distributes along the z-axis corresponding to the p_z-like orbital, the light hole moves along the z-axis more easily than on the xy-plane, which results in $m_{z,\text{lh}}{}^* < m_{xy,\text{lh}}{}^*$. For example, in GaAs, we have $m_{z,\text{hh}}{}^* = 0.377\, m$, $m_{z,\text{lh}}{}^* = 0.09\, m$, $m_{xy,\text{hh}}{}^* = 0.11\, m$, and $m_{xy,\text{lh}}{}^* = 0.21\, m$, where m is the electron mass in a vacuum.

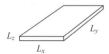

Fig. 7.4. Quantum well

(i) Wave Function and Momentum Matrix Element

We assume that a wave vector $\boldsymbol{k}$ is directed toward the z-axis and contribution of the split-off band is negligible. Hence, we consider the heavy hole band and the light hole band as the valence bands. If we express the up-spin and down-spin as α and β, respectively, the wave functions of the conduction band are written as

$$|s\alpha\rangle, \quad |s\beta\rangle. \tag{7.10}$$

Quantum states of the valence bands are indicated by $\boldsymbol{j}$, which is a sum of the angular momentum operator $\boldsymbol{l}$ and the spin operator $\boldsymbol{s}$ when the spin-orbit interaction is considered. We can express the wave function $|j, m_j\rangle$ as

For the heavy hole

$$\left|\frac{3}{2}, \frac{3}{2}\right\rangle = \frac{1}{\sqrt{2}}|(x + \mathrm{i}\,y)\alpha\rangle,$$

$$\left|\frac{3}{2}, -\frac{3}{2}\right\rangle = \frac{1}{\sqrt{2}}|(x - \mathrm{i}\,y)\beta\rangle, \tag{7.11}$$

For the light hole

$$\left|\frac{3}{2}, \frac{1}{2}\right\rangle = \frac{1}{\sqrt{6}}|2z\alpha + (x + \mathrm{i}\,y)\beta\rangle,$$

$$\left|\frac{3}{2}, -\frac{1}{2}\right\rangle = \frac{1}{\sqrt{6}}|2z\beta - (x - \mathrm{i}\,y)\alpha\rangle,$$

(7.12)

where j is the eigenvalue of $\boldsymbol{j}$ and m_j is the eigenvalue of j_z.

In the following, we will consider the optical transitions between the conduction band and the valence bands. As shown in (2.41), the optical power gain coefficient g is in proportion to a square of the momentum matrix element $\langle 1|\boldsymbol{p}|2\rangle^2$. Therefore, to examine the anisotropic optical gain, we compare the momentum matrix elements for the optical transitions.

Along each axis, the momentum matrix elements between the conduction band and the heavy hole band are given by

$$x\text{-axis: } \frac{1}{\sqrt{2}}\sqrt{3}M,$$

$$y\text{-axis: } \pm\mathrm{i}\,\frac{1}{\sqrt{2}}\sqrt{3}M,$$

$$z\text{-axis: } 0.$$

Here, from (1.6) and (1.42), M is defined as

$$\sqrt{3}M \equiv \langle s|p_x|x\rangle = \langle s|p_y|y\rangle = \langle s|p_z|z\rangle$$

$$= m\left[\frac{1}{2m_\mathrm{e}^*}\frac{E_\mathrm{g}(E_\mathrm{g} + \Delta_0)}{E_\mathrm{g} + \frac{2}{3}\Delta_0}\right]^{1/2},$$

(7.13)

where m is the electron mass in a vacuum, m_e^* is the effective mass of the electron in the conduction band, E_g is the bandgap energy, and Δ_0 is the split-off energy due to the spin-orbit interaction. A coefficient $\sqrt{3}$ is introduced so that a matrix element averaged over all directions of the wave vector $\boldsymbol{k}$ may be M.

(ii) Optical Transitions in QWs

As shown in Fig. 7.5, we assume that a quantization axis is the z-axis and a QW layer is placed on the xy-plane. If a light is supposed to propagate along a positive direction of the x-axis, an electric field E along the y-axis is the TE mode, and an electric field E along the z-axis is the TM mode. Because the wave vector $\boldsymbol{k}$ can take various directions, we express a direction of $\boldsymbol{k}$ in a polar coordinate system, as shown in Fig. 7.5. In this QW, the z-component of $\boldsymbol{k}$ is discrete, whereas the x- and y-components of $\boldsymbol{k}$ are continuous. For example, if the energy barrier is infinite, we have $k_z = n\pi/L_z$ with an integer n.

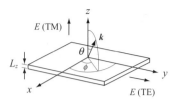

Fig. 7.5. Direction of a wave vector $\boldsymbol{k}$

According to the directions of $\boldsymbol{k}$, the momentum matrix elements between the conduction band and the heavy hole band are given by

$$x\text{-axis:}\quad \frac{1}{\sqrt{2}}\sqrt{3}M(\cos\theta\cos\phi \mp \mathrm{i}\sin\phi),$$

$$y\text{-axis:}\quad \frac{1}{\sqrt{2}}\sqrt{3}M(\cos\theta\sin\phi \pm \mathrm{i}\cos\phi),$$

$$z\text{-axis:}\quad -\frac{1}{\sqrt{2}}\sqrt{3}M\sin\theta.$$

The square of the optical transition matrix elements is proportional to $\langle 1|\boldsymbol{E}\cdot\boldsymbol{p}|2\rangle^2$, where $\boldsymbol{E}$ is the electric field. Therefore, only the momentum matrix element with a component parallel to $\boldsymbol{E}$ contributes to the optical transitions. We consider a wave vector $\boldsymbol{k}_n$ for a quantum number n, and we average the square of the momentum matrix element all over the directions on the xy-plane by fixing the z-component k_{nz} of $\boldsymbol{k}_n$.

(iii) Momentum Matrix Elements between the Conduction Band and the Heavy Hole Band in QWs

An average of the squared momentum matrix element $\langle M^2\rangle_{\mathrm{hh,TE}}$ for the TE mode $(\boldsymbol{E}/\!/y)$ is expressed as

$$\langle M^2\rangle_{\mathrm{hh,TE}} = \frac{3M^2}{2}\cdot\frac{1}{2\pi}\int_0^{2\pi}(\cos^2\theta\sin^2\phi + \cos^2\phi)\,\mathrm{d}\phi$$

$$= \frac{3M^2}{4}(1+\cos^2\theta) = \frac{3M^2}{4}\left(1+\frac{k_z{}^2}{k^2}\right)$$

$$= \frac{3M^2}{4}\left(1+\frac{E_{z,n}}{E_n}\right), \tag{7.14}$$

where $E_{z,n}$ is the quantized energy of the nth subband and E_n is the total energy of the nth subband.

At the subband edge, $E_{z,n} = E_n$ is satisfied; then we have

$$\langle M^2\rangle_{\mathrm{hh,TE}} = \frac{3M^2}{2}. \tag{7.15}$$

An average of the squared momentum matrix element $\langle M^2 \rangle_{\text{hh,TM}}$ for the TM mode ($\boldsymbol{E}//z$) is given by

$$\langle M^2 \rangle_{\text{hh,TM}} = \frac{3M^2}{2} \sin^2\theta = \frac{3M^2}{2} (1 - \cos^2\theta)$$
$$= \frac{3M^2}{2} \left(1 - \frac{k_z^2}{k^2}\right) = \frac{3M^2}{2} \left(1 - \frac{E_{z,n}}{E_n}\right). \tag{7.16}$$

As a result, at the subband edge, we obtain

$$\langle M^2 \rangle_{\text{hh,TM}} = 0. \tag{7.17}$$

As shown in (7.15) and (7.17), there is a *selection rule* between the conduction band and the heavy hole band at the subband edge, that is, only the optical transition for the TE mode is allowed; that for the TM mode is inhibited.

(iv) Momentum Matrix Elements between the Conduction Band and the Light Hole Band in QWs

An average of the squared momentum matrix element $\langle M^2 \rangle_{\text{lh,TE}}$ for the TE mode ($\boldsymbol{E}//y$) is obtained as

$$\langle M^2 \rangle_{\text{lh,TE}} = \frac{M^2}{4} (1 + \cos^2\theta) + M^2 \sin^2\theta$$
$$= \frac{M^2}{4} \left(1 + \frac{E_{z,n}}{E_n}\right) + M^2 \left(1 - \frac{E_{z,n}}{E_n}\right). \tag{7.18}$$

Hence, at the subband edge, we have

$$\langle M^2 \rangle_{\text{lh,TE}} = \frac{M^2}{2}. \tag{7.19}$$

An average of the squared momentum matrix element $\langle M^2 \rangle_{\text{lh,TM}}$ for the TM mode ($\boldsymbol{E}//z$) is written as

$$\langle M^2 \rangle_{\text{lh,TM}} = \frac{M^2}{2} \sin^2\theta + 2M^2 \cos^2\theta$$
$$= \frac{M^2}{2} \left(1 - \frac{E_{z,n}}{E_n}\right) + 2M^2 \frac{E_{z,n}}{E_n}. \tag{7.20}$$

Therefore, at the subband edge, we obtain

$$\langle M^2 \rangle_{\text{lh,TM}} = 2M^2. \tag{7.21}$$

(v) Optical Gains in QW-LDs

In Fig. 7.3, the vertical line shows the electron energy, and the hole energy decreases with an increase in the height. As a result, the concentration of the heavy hole is larger than that of the light hole. Thus, among the polarization-dependent optical gain $g_{e\text{-}hh,TE}$, $g_{e\text{-}lh,TE}$, and $g_{e\text{-}lh,TM}$, we have the following relation

$$g_{e\text{-}hh,TE} > g_{e\text{-}lh,TM} > g_{e\text{-}lh,TE},$$

where subscripts e-hh and e-lh show the recombination of the electron and the heavy hole and of the electron and the light hole, respectively. As shown earlier, the optical gain for the TE mode has a maximum value in the optical transition between the conduction band and the heavy hole band in the QW-LDs.

(d) Low-Loss Optical Waveguides

In the QWs, the densities of states are step functions, and the absorption coefficient at the band edge changes sharply with wavelength. Therefore, the absorption loss in the optical waveguides with the QWs is lower than that with the bulk DHs.

(e) High-Speed Modulation

As shown in (5.103), the relaxation oscillation frequency f_r is given by

$$f_r = \frac{1}{2\pi}\sqrt{\frac{\partial G}{\partial n}\frac{S_0}{\tau_{ph}}} . \tag{7.22}$$

In the QW-LDs, based on the step-like densities of states, the differential optical gain $\partial G/\partial n$ is larger than that in the bulk DH-LDs. Hence, f_r in the QW-LDs is larger than that in the bulk DH-LDs, which leads to high-speed modulations in the QW-LDs. Figure 7.6 shows a calculated differential optical gain $(n_r/c)\partial G/\partial n$ as a function of the quasi-Fermi level of the conduction band E_{Fc}. It is clearly revealed that the differential optical gain in the QW-LDs is larger than that in the bulk DH-LDs.

From (7.6), the optical gain is simplified as

$$g = |M|^2_{ave}\,\rho\,[f_c - f_v]. \tag{7.23}$$

The QWs modify the density of states ρ in (7.23), while the intentionally doped active layers can change $[f_c - f_v]$. For example, p-doped active layers reduce f_v, and a relaxation oscillation frequency of about 30 GHz has been reported.

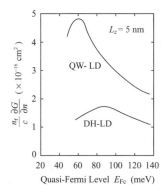

Fig. 7.6. Differential optical gain

(f) Narrow Spectral Linewidth

From (5.169), the spectral linewidth $\Delta\omega_0$ of a laser light is expressed as

$$\Delta\omega_0 = \frac{\hbar\omega_m c E_{cv} \ln(1/R)}{4P_0 n_r L}(1+\alpha^2), \tag{7.24}$$

where the spectral linewidth enhancement factor α is given by

$$\alpha = -\frac{2\omega_m}{n_r}\frac{\partial n_r}{\partial n}\left(\frac{\partial G}{\partial n}\right)^{-1}, \tag{7.25}$$

as shown in (5.179). Here, n_r is the refractive index, n is the carrier concentration, and G is the amplification rate. As illustrated in Fig. 7.6, the differential optical gain $\partial G/\partial n$ in the QW-LDs is larger than that in the bulk DH-LDs. As a result, from (7.25), the absolute value of α in the QW-LDs is small, which leads to a narrow spectral linewidth $\Delta\omega_0$. Figure 7.7 shows calculated α values as a function of the quasi-Fermi level of the conduction band E_{Fc}. It is found that α in the QW-LDs is smaller than that in the bulk DH-LDs. Moreover, a small α leads to low chirping during modulation, which is suitable for long-haul, large-capacity optical fiber communication systems.

7.3 Strained Quantum Well LDs

7.3.1 Effect of Strains

According to *group theory*, with a decrease in symmetry of the crystals, degeneracy of the energy band is removed. In the bulk structures for semiconductor lasers, due to their high symmetry, the heavy hole band and the light hole band are degenerate at Γ point ($\boldsymbol{k} = 0$). In the QWs, due to their lower

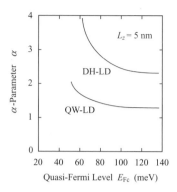

Fig. 7.7. Spectral linewidth enhancement factor

symmetry than that of the bulk structures, degeneracy at $k = 0$ is removed, as shown in Fig. 7.3.

When the *strains* are applied to semiconductor crystals, symmetry of the crystals is reduced, and degeneracy of the energy bands is removed. According to the compressive or tensile strains, the band structures change in different ways. Based on this principle, *band-structure engineering*, which modifies the band structures by intentionally controlling the strains, has been developed. By introducing the strains into the active layers, we can obtain excellent characteristics, such as a low threshold current, a high differential quantum efficiency, high-speed modulation, low chirping, and a narrow spectral linewidth in semiconductor lasers, as in the QW-LDs. With regard to the optical gain for the TE mode g_{TE} and that for the TM mode g_{TM}, the compressive strains in the active layer plane lead to $g_{TE} > g_{TM}$, while the tensile strains result in $g_{TE} < g_{TM}$.

To induce the strains in semiconductor crystals, an external *stress*, a difference in the thermal expansion coefficients of the materials, and a difference in the lattice constants are used. Especially, a difference in the lattice constants, which is referred to as *lattice mismatching*, is frequently adopted, because the most stable strains are obtained. The lattice mismatching takes place when a semiconductor layer is epitaxially grown on the substrate with a different lattice constant from that of the grown layer. If thickness of the grown layer exceeds the *critical thickness*, the dislocations are generated in the grown layer. If the layer thickness further increases up to about $1\,\mu\text{m}$, which is much larger than the critical thickness, the dislocations are sometimes reduced, and such a thick grown layer can be used as the *buffer layer*. However, this thick layer is not suitable for active layers, because the threshold current is large. When the grown layer is thinner than the critical thickness as in the QWs, the dislocations are not generated. Hence, the *strained QWs* are suitable for band-structure engineering.

7.3.2 Band-Structure Engineering

(a) Epitaxial Growth and Strains

When the epitaxially grown layers are thinner than the critical thickness, the epitaxial layers are grown on the substrate with their lattice constants matched to that of the substrate. Therefore, the *elastic strains* are induced in the epitaxial layers. We assume that the lattice constant of an epitaxial layer itself is $a(x)$ and the lattice constant of the substrate is a_0. For $a(x) > a_0$, the *compressive strain* is generated in the grown layer, while for $a(x) < a_0$, the *tensile strain* is produced in the grown layer.

The strain *tensors* are expressed by matrixes using their symmetrical properties. The matrix elements of the strains ε_{ij} consist of the *diagonal elements* ε_{ii} called the *hydrostatic strains* and the *nondiagonal elements* ε_{ij} $(i \neq j)$ named the *shear strains*.

(b) Low Threshold Current

The radii of the curvature of the energy bands, that is, the effective masses of the carriers, are modified by the strains. With decreases in the effective masses of the carriers, the carrier concentration to achieve the population inversion decreases. Therefore, a low threshold current is obtained, as in the following.

When the electric currents are injected in semiconductor lasers, many electrons are generated in the conduction band, and many holes are simultaneously produced in the valence bands. In this case, the carrier distributions are far from thermal equilibrium. Therefore, we cannot express the distribution functions of the electron and the hole with a single Fermi level, E_F. As a result, regarding the electrons in the conduction band and the holes in the valence band independently take the Fermi-Dirac distributions, and we introduce the quasi-Fermi levels E_{Fc} and E_{Fv}, which are defined as

$$n \equiv N_c \exp\left(-\frac{E_c - E_{Fc}}{k_B T}\right),$$

$$p \equiv N_v \exp\left(-\frac{E_{Fv} - E_v}{k_B T}\right),$$

$$N_c = 2\left(\frac{2\pi m_e^* k_B T}{h^2}\right)^{3/2},$$

$$N_v = 2\left(\frac{2\pi m_h^* k_B T}{h^2}\right)^{3/2},$$

$$(7.26)$$

where n and p are the concentrations of the electron and the hole, respectively; E_c and E_v are the energies of the bottom of the conduction band and the top of the valence band, respectively; k_B is the Boltzmann constant; T is absolute temperature; N_c and N_v are the effective densities of states for the electron and the hole, respectively; $m_e{}^*$ and $m_h{}^*$ are the effective masses of the electron and the hole, respectively; and h is Planck's constant.

From (7.26), E_{Fc} and E_{Fv} are expressed as

$$E_{Fc} = E_c + k_B T \ln \left(\frac{n}{N_c} \right),$$

$$E_{Fv} = E_v - k_B T \ln \left(\frac{p}{N_v} \right). \tag{7.27}$$

Hence, we obtain

$$E_{Fc} - E_{Fv} = E_g + k_B T \ln \frac{np}{N_c N_v}, \quad E_g = E_c - E_v. \tag{7.28}$$

As shown in (2.29), the condition for the population inversion is given by

$$E_{Fc} - E_{Fv} > E_g, \tag{7.29}$$

which reduces to

$$np > N_c N_v = 4 \left(\frac{2\pi k_B T}{h^2} \right)^3 (m_e{}^* m_h{}^*)^{3/2}, \tag{7.30}$$

where (7.26) and (7.28) were used. From (7.30), the transparent carrier concentration n_0 is obtained as

$$n_0 = 2 \left(\frac{2\pi k_B T}{h^2} \right)^{3/2} (m_e{}^* m_h{}^*)^{3/4}. \tag{7.31}$$

Therefore, to achieve a low n_0 leading to a low threshold, $m_e{}^*$ and $m_h{}^*$ should be light.

As shown in (1.11), the effective mass tensor m_{ij} of the carrier is written as

$$\left(\frac{1}{m} \right)_{ij} = \frac{1}{\hbar^2} \frac{\partial^2 E}{\partial k_i \partial k_j}, \tag{7.32}$$

which is related to the radius of the curvature of the band energy. The compressive strains in the active layer decrease the radius of the curvature of the heavy hole band in the vicinity of the band edge, which reduces the effective mass of the heavy hole. As a result, the threshold carrier concentration decreases, and therefore the Auger recombination rate in (5.52) is also reduced. Furthermore, from the momentum conservation law and the energy conservation law, the Auger transitions are going to be inhibited. Hence, the Auger processes are drastically reduced. As a result, a high light emission efficiency and a low threshold current density are simultaneously obtained. The strains in the active layers also enhance the differential optical gain $\partial G / \partial n$, which further reduces the threshold carrier concentration.

(c) Anisotropic Optical Gain

The tensile strains make the energy of the light hole lower than that of the heavy hole. When we draw the energy bands with the vertical line as the energy of the electron, the band edge of the light hole is above that of the heavy hole. As a result, concentration of the light hole is larger than that of the heavy hole, and the optical gain for the TM mode is larger than that for the TE mode, which results in laser oscillation in the TM mode. By optimizing the tensile strains so that the optical gain for the TE mode and that for the TM mode may take a common value, polarization-independent semiconductor optical amplifiers have been demonstrated.

(d) High-Speed Modulation

The strains in the active layers enhance the differential optical gain $\partial G/\partial n$, which leads to high-speed modulations. The compressive strains result in large gain saturation, while the tensile strains lead to low gain saturation with a large optical gain. Hence, the tensile strains are expected to improve high-speed modulation characteristics.

(e) Narrow Spectral Linewidth

The strains increase the differential optical gain $\partial G/\partial n$ and decrease α. Therefore, a narrow spectral linewidth such as 3.6 kHz has been obtained in a strained QW $\lambda/4$-shifted DFB-LD [49].

7.3.3 Analysis

(a) Fundamental Equations

We introduce the effect of strains into the Schrödinger equation as the perturbation, where the unperturbed Hamiltonian is assumed to include the $\boldsymbol{k}\cdot\boldsymbol{p}$ perturbation and the spin-orbit interaction. This unperturbed Hamiltonian whose base wave function is $|j, m_j\rangle$ is known as *Luttinger-Kohn Hamiltonian* [50], while the perturbation Hamiltonian representing the effect of the strains is called *Pikus-Bir Hamiltonian* [51–53].

(i) Luttinger-Kohn Hamiltonian

In Chapter 1, to obtain the eigenenergies of the valence bands, we used the following Hamiltonian

$$\mathcal{H} = \mathcal{H}_0 + \mathcal{H}_{kp} + \mathcal{H}_{\mathrm{so}}, \tag{7.33}$$

where $\mathcal{H}_0$ is the unperturbed Hamiltonian, $\mathcal{H}_{kp}$ is the $\boldsymbol{k} \cdot \boldsymbol{p}$ perturbation Hamiltonian, and $\mathcal{H}_{so}$ is the spin-orbit interaction Hamiltonian. Using this Hamiltonian $\mathcal{H}$, we consider a 6×6 matrix expressed as

$$
\begin{bmatrix}
\mathcal{H}_{11} & \mathcal{H}_{12} & \mathcal{H}_{13} & \mathcal{H}_{14} & \mathcal{H}_{15} & \mathcal{H}_{16} \\
\mathcal{H}_{21} & \mathcal{H}_{22} & \mathcal{H}_{23} & \mathcal{H}_{24} & \mathcal{H}_{25} & \mathcal{H}_{26} \\
\mathcal{H}_{31} & \mathcal{H}_{32} & \mathcal{H}_{33} & \mathcal{H}_{34} & \mathcal{H}_{35} & \mathcal{H}_{36} \\
\mathcal{H}_{41} & \mathcal{H}_{42} & \mathcal{H}_{43} & \mathcal{H}_{44} & \mathcal{H}_{45} & \mathcal{H}_{46} \\
\mathcal{H}_{51} & \mathcal{H}_{52} & \mathcal{H}_{53} & \mathcal{H}_{54} & \mathcal{H}_{55} & \mathcal{H}_{56} \\
\mathcal{H}_{61} & \mathcal{H}_{62} & \mathcal{H}_{63} & \mathcal{H}_{24} & \mathcal{H}_{65} & \mathcal{H}_{66}
\end{bmatrix}, \tag{7.34}
$$

where $\mathcal{H}_{ij} = \langle i|\mathcal{H}|j \rangle$ and $\langle i|$ and $|j \rangle$ are written as

$$\langle 1| = \left\langle \tfrac{3}{2}, \tfrac{3}{2} \right|, \quad \langle 2| = \left\langle \tfrac{3}{2}, \tfrac{1}{2} \right|, \quad \langle 3| = \left\langle \tfrac{3}{2}, -\tfrac{1}{2} \right|, \quad \langle 4| = \left\langle \tfrac{3}{2}, -\tfrac{3}{2} \right|,$$

$$\langle 5| = \left\langle \tfrac{1}{2}, \tfrac{1}{2} \right|, \quad \langle 6| = \left\langle \tfrac{1}{2}, -\tfrac{1}{2} \right|, \; |1\rangle = \left| \tfrac{3}{2}, \tfrac{3}{2} \right\rangle, \quad |2\rangle = \left| \tfrac{3}{2}, \tfrac{1}{2} \right\rangle, \tag{7.35}$$

$$|3\rangle = \left| \tfrac{3}{2}, -\tfrac{1}{2} \right\rangle, \; |4\rangle = \left| \tfrac{3}{2}, -\tfrac{3}{2} \right\rangle, \; |5\rangle = \left| \tfrac{1}{2}, \tfrac{1}{2} \right\rangle, \; |6\rangle = \left| \tfrac{1}{2}, -\tfrac{1}{2} \right\rangle.$$

Equation (7.34) shows Luttinger-Kohn Hamiltonian for the valence bands, and this Hamiltonian is used as the unperturbed Hamiltonian to analyze the strains. When the split-off energy Δ_0 is larger than $0.3\,\mathrm{eV}$ as in GaAs, contribution of the split-off band can be neglected. In this case, Luttinger-Kohn Hamiltonian for the valence bands $\mathcal{H}_{LK}$ results in a 4×4 matrix, which is given by

$$
\mathcal{H}_{LK} = \begin{bmatrix}
\mathcal{H}_{11} & \mathcal{H}_{12} & \mathcal{H}_{13} & \mathcal{H}_{14} \\
\mathcal{H}_{21} & \mathcal{H}_{22} & \mathcal{H}_{23} & \mathcal{H}_{24} \\
\mathcal{H}_{31} & \mathcal{H}_{32} & \mathcal{H}_{33} & \mathcal{H}_{34} \\
\mathcal{H}_{41} & \mathcal{H}_{42} & \mathcal{H}_{43} & \mathcal{H}_{44}
\end{bmatrix} = \begin{bmatrix}
a_+ & b & c & 0 \\
b^* & a_- & 0 & c \\
c^* & 0 & a_- & -b \\
0 & c^* & -b^* & a_+
\end{bmatrix}, \tag{7.36}
$$

where

$$a_\pm = \frac{\hbar^2}{2m} \left[-(\gamma_1 \mp 2\gamma_2)k_z{}^2 - (\gamma_1 \pm \gamma_2)(k_x{}^2 + k_y{}^2) \right],$$

$$b = \frac{\hbar^2}{m} \sqrt{3}\, \gamma_3 (k_x - \mathrm{i}\, k_y)k_z, \tag{7.37}$$

$$c = \frac{\hbar^2}{2m} \sqrt{3} \left[\gamma_2 (k_x{}^2 - k_y{}^2) - \mathrm{i}\, 2\gamma_3 k_x k_y \right].$$

Here, γ_1, γ_2, and γ_3 are *Luttinger parameters*, and they are related to the effective masses $m_{hh}{}^*$ and $m_{lh}{}^*$ of the heavy hole and the light hole as

$$\frac{1}{m} (\gamma_1 - 2\gamma_2) = \frac{1}{m_{hh}{}^*} \quad \text{(heavy hole)}, \tag{7.38}$$

$$\frac{1}{m} (\gamma_1 + 2\gamma_2) = \frac{1}{m_{lh}{}^*} \quad \text{(light hole)}, \tag{7.39}$$

where m is the electron mass in a vacuum.

We can also express (7.36) as

$$
\mathcal{H}_{\mathrm{LK}} = \frac{\hbar^2}{2m} \left[-\left(\gamma_1 + \frac{5}{2}\gamma_2 \right) k^2 + 2\gamma_2 (k_x{}^2 J_x{}^2 + k_y{}^2 J_y{}^2 + k_z{}^2 J_z{}^2) \right.
$$
$$
\left. + 4\gamma_3 \left[\{k_x k_y\}\{J_x J_y\} + \{k_y k_z\}\{J_y J_z\} + \{k_z k_x\}\{J_z J_x\} \right] \right], \quad (7.40)
$$

where J_x, J_y, and J_z are the matrixes represented as

$$
J_x = \frac{1}{2}
\begin{bmatrix}
0 & \sqrt{3}\,i & 0 & 0 \\
-\sqrt{3}\,i & 0 & 2\,i & 0 \\
0 & -2\,i & 0 & \sqrt{3}\,i \\
0 & 0 & -\sqrt{3}\,i & 0
\end{bmatrix},
$$

$$
J_y = \frac{1}{2}
\begin{bmatrix}
0 & \sqrt{3} & 0 & 0 \\
\sqrt{3} & 0 & 2 & 0 \\
0 & 2 & 0 & \sqrt{3} \\
0 & 0 & \sqrt{3} & 0
\end{bmatrix}, \quad (7.41)
$$

$$
J_z = \frac{1}{2}
\begin{bmatrix}
3 & 0 & 0 & 0 \\
0 & 1 & 0 & 0 \\
0 & 0 & -1 & 0 \\
0 & 0 & 0 & -3
\end{bmatrix}.
$$

In (7.40), $\{k_x k_y\}$ and $\{J_x J_y\}$ are defined as

$$
\{k_x k_y\} \equiv \frac{1}{2}(k_x k_y + k_y k_x), \quad \{J_x J_y\} \equiv \frac{1}{2}(J_x J_y + J_y J_x). \quad (7.42)
$$

(ii) Pikus-Bir Hamiltonian

Pikus-Bir Hamiltonian $\mathcal{H}_s$ is given by a sum of the *orbit-strain interaction Hamiltonian* $\mathcal{H}_{\mathrm{os}}$ and the *strain-dependent spin-orbit interaction Hamiltonian* $\mathcal{H}_{\mathrm{ss}}$. In the valence bands of the zinc-blende structure, the orbit-strain interaction Hamiltonian $\mathcal{H}_{\mathrm{os}}$ at Γ point $(\boldsymbol{k} = 0)$ is written as

$$
\mathcal{H}_{\mathrm{os}} = -a_1(\varepsilon_{xx} + \varepsilon_{yy} + \varepsilon_{zz})
$$
$$
- 3b_1 \left[\left(L_x{}^2 - \frac{\boldsymbol{L}^2}{3} \right)\varepsilon_{xx} + \left(L_y{}^2 - \frac{\boldsymbol{L}^2}{3} \right)\varepsilon_{yy} + \left(L_z{}^2 - \frac{\boldsymbol{L}^2}{3} \right)\varepsilon_{zz} \right]
$$
$$
- \sqrt{3}d_1 \left(L_x L_y + L_y L_x \right)\varepsilon_{xy} - \sqrt{3}d_1 \left(L_y L_z + L_z L_y \right)\varepsilon_{yz}
$$
$$
- \sqrt{3}d_1 \left(L_z L_x + L_x L_z \right)\varepsilon_{zx}. \quad (7.43)
$$

The strain-dependent spin-orbit interaction Hamiltonian $\mathcal{H}_{\mathrm{ss}}$ is expressed as

$$\mathcal{H}_{ss} = -a_2(\varepsilon_{xx} + \varepsilon_{yy} + \varepsilon_{zz})(\boldsymbol{L} \cdot \boldsymbol{s})$$
$$- 3b_2 \left(L_x s_x - \frac{\boldsymbol{L} \cdot \boldsymbol{s}}{3} \right) \varepsilon_{xx} - 3b_2 \left(L_y s_y - \frac{\boldsymbol{L} \cdot \boldsymbol{s}}{3} \right) \varepsilon_{yy}$$
$$- 3b_2 \left(L_z s_z - \frac{\boldsymbol{L} \cdot \boldsymbol{s}}{3} \right) \varepsilon_{zz}$$
$$- \sqrt{3}d_2 \left(L_x s_y + L_y s_x \right) \varepsilon_{xy} - \sqrt{3}d_2 \left(L_y s_z + s_z L_y \right) \varepsilon_{yz}$$
$$- \sqrt{3}d_2 \left(L_z s_x + L_x s_z \right) \varepsilon_{zx}, \tag{7.44}$$

where L_x, L_y, L_z, and $\boldsymbol{L}$ are the *orbital angular momentum operators*; s_x, s_y, s_z, and $\boldsymbol{s}$ are the *spin angular momentum operators*; $\varepsilon_{ij}(i, j = x, y, z)$ is a matrix element for the strain tensor; and a_i, b_i, and d_i $(i = 1, 2)$ are the *deformation potentials*.

Because the orbit-strain interaction Hamiltonian $\mathcal{H}_{os}$ is larger than the strain-dependent spin-orbit interaction Hamiltonian $\mathcal{H}_{ss}$, the strained QWs are often analyzed by considering only $\mathcal{H}_{os}$.

(iii) Relationship between Strain and Stress

The strains ε and stresses σ are both tensors, which are related by

$$\sigma_{ij} = \sum_{k,l} c_{ijkl} \varepsilon_{kl} = c_{ijkl} \varepsilon_{kl}, \tag{7.45}$$

where c_{ijkl} is the elastic stiffness constant. Also, on the right-hand side, $\sum$ is omitted by promising that we take a sum with regard to a subscript appearing twice, which is referred to as the *Einstein summation convention*.

It is useful to express the tensors by the matrixes, because of symmetry in the tensors. As a coordinate system for the tensors, we use 1, 2, and 3 as indexes. For example, 1, 2, and 3 correspond to x, y, and z, respectively. Here, we relate indexes 1, 2, and 3 of the tensors to indexes 1, 2, 3, 4, 5, and 6 of the matrixes as follows:

Tensor expression	11	22	33	23, 32	31, 13	12, 21
Matrix element	1	2	3	4	5	6

Therefore, we can rewrite (7.45) as

$$\sigma_i = \sum_j c_{ij} \varepsilon_j = c_{ij} \varepsilon_j \quad (i, j = 1, 2, 3, 4, 5, 6), \tag{7.46}$$

where

$$\begin{bmatrix} \sigma_{11} & \sigma_{12} & \sigma_{13} \\ \sigma_{21} & \sigma_{22} & \sigma_{23} \\ \sigma_{31} & \sigma_{32} & \sigma_{33} \end{bmatrix} = \begin{bmatrix} \sigma_{11} & \sigma_{12} & \sigma_{31} \\ \sigma_{12} & \sigma_{22} & \sigma_{23} \\ \sigma_{31} & \sigma_{23} & \sigma_{33} \end{bmatrix} = \begin{bmatrix} \sigma_1 & \sigma_6 & \sigma_5 \\ \sigma_6 & \sigma_2 & \sigma_4 \\ \sigma_5 & \sigma_4 & \sigma_3 \end{bmatrix}, \tag{7.47}$$

$$
\begin{bmatrix} \varepsilon_{11} & \varepsilon_{12} & \varepsilon_{13} \\ \varepsilon_{21} & \varepsilon_{22} & \varepsilon_{23} \\ \varepsilon_{31} & \varepsilon_{32} & \varepsilon_{33} \end{bmatrix} = \begin{bmatrix} \varepsilon_{11} & \varepsilon_{12} & \varepsilon_{31} \\ \varepsilon_{12} & \varepsilon_{22} & \varepsilon_{23} \\ \varepsilon_{31} & \varepsilon_{23} & \varepsilon_{33} \end{bmatrix} = \frac{1}{2} \begin{bmatrix} 2\varepsilon_1 & \varepsilon_6 & \varepsilon_5 \\ \varepsilon_6 & 2\varepsilon_2 & \varepsilon_4 \\ \varepsilon_5 & \varepsilon_4 & 2\varepsilon_3 \end{bmatrix}. \tag{7.48}
$$

In the zinc-blende structures with symmetry of $\bar{4}3m$ or T_d, (7.45) reduces to [54, 55]

$$
\begin{bmatrix} \sigma_1 \\ \sigma_2 \\ \sigma_3 \\ \sigma_4 \\ \sigma_5 \\ \sigma_6 \end{bmatrix} = \begin{bmatrix} c_{11} & c_{12} & c_{12} & 0 & 0 & 0 \\ c_{12} & c_{11} & c_{12} & 0 & 0 & 0 \\ c_{12} & c_{12} & c_{11} & 0 & 0 & 0 \\ 0 & 0 & 0 & c_{44} & 0 & 0 \\ 0 & 0 & 0 & 0 & c_{44} & 0 \\ 0 & 0 & 0 & 0 & 0 & c_{44} \end{bmatrix} \begin{bmatrix} \varepsilon_1 \\ \varepsilon_2 \\ \varepsilon_3 \\ \varepsilon_4 \\ \varepsilon_5 \\ \varepsilon_6 \end{bmatrix}. \tag{7.49}
$$

(b) Bulk Structures

Here, we consider a semiconductor crystal in which *biaxial stresses* are applied to the crystal plane due to the lattice mismatching. We assume that the epitaxial layers are grown along the z-axis, and the layer plane is on the xy-plane. When the lattice constant of the substrate is a_0 and that of the epitaxially grown layer itself is $a(x)$, the strains due to lattice mismatching are given by

$$
\varepsilon_{xx} = \varepsilon_{yy} = \frac{a_0 - a(x)}{a(x)} = \varepsilon, \quad \varepsilon_{zz} \neq 0,
$$
$$
\tag{7.50}
$$
$$
\varepsilon_{xy} = \varepsilon_{yz} = \varepsilon_{zx} = 0,
$$

where $\varepsilon < 0$ corresponds to the compressive strain and $\varepsilon > 0$ shows the tensile strain.

When the biaxial stresses are induced in the layer plane (xy-plane) due to lattice mismatching, neither the stress along the growth axis (z-axis) nor the shear stresses are imposed to the epitaxial layer. Therefore, the biaxial stresses are expressed as

$$
\sigma_{xx} = \sigma_{yy} = \sigma, \quad \sigma_{zz} = 0,
$$
$$
\tag{7.51}
$$
$$
\sigma_{xy} = \sigma_{yz} = \sigma_{zx} = 0.
$$

If we relate indexes 1, 2, and 3 of the tensors to x, y, and z, respectively, and substitute (7.50) and (7.51) into (7.49), we have

$$
\begin{bmatrix} \sigma \\ \sigma \\ 0 \\ 0 \\ 0 \\ 0 \end{bmatrix} = \begin{bmatrix} c_{11} & c_{12} & c_{12} & 0 & 0 & 0 \\ c_{12} & c_{11} & c_{12} & 0 & 0 & 0 \\ c_{12} & c_{12} & c_{11} & 0 & 0 & 0 \\ 0 & 0 & 0 & c_{44} & 0 & 0 \\ 0 & 0 & 0 & 0 & c_{44} & 0 \\ 0 & 0 & 0 & 0 & 0 & c_{44} \end{bmatrix} \begin{bmatrix} \varepsilon \\ \varepsilon \\ \varepsilon_{zz} \\ 0 \\ 0 \\ 0 \end{bmatrix}. \tag{7.52}
$$

From (7.52), we obtain

$$\sigma = (c_{11} + c_{12})\varepsilon + c_{12}\varepsilon_{zz},$$

$$0 = 2c_{12}\varepsilon + c_{11}\varepsilon_{zz},$$

(7.53)

which leads to

$$\varepsilon_{zz} = -\frac{2c_{12}}{c_{11}}\varepsilon,$$

$$\sigma = \left(c_{11} + c_{12} - \frac{2c_{12}{}^2}{c_{11}}\right)\varepsilon.$$

(7.54)

Substituting (7.54) into (7.43) results in

$$\mathcal{H}_{os} = -2a_1\left(1 - \frac{c_{12}}{c_{11}}\right)\varepsilon + 3b_1\left(L_z{}^2 - \frac{1}{3}L^2\right)\left(1 + \frac{2c_{12}}{c_{11}}\right)\varepsilon.$$

(7.55)

Here, we consider a matrix $\mathcal{H}_{sv}$ such as

$$
\mathcal{H}_{sv} = \begin{bmatrix}
\langle 1|\mathcal{H}_{os}|1\rangle & \langle 1|\mathcal{H}_{os}|2\rangle & \langle 1|\mathcal{H}_{os}|3\rangle & \langle 1|\mathcal{H}_{os}|4\rangle \\
\langle 2|\mathcal{H}_{os}|1\rangle & \langle 2|\mathcal{H}_{os}|2\rangle & \langle 2|\mathcal{H}_{os}|3\rangle & \langle 2|\mathcal{H}_{os}|4\rangle \\
\langle 3|\mathcal{H}_{os}|1\rangle & \langle 3|\mathcal{H}_{os}|2\rangle & \langle 3|\mathcal{H}_{os}|3\rangle & \langle 3|\mathcal{H}_{os}|4\rangle \\
\langle 4|\mathcal{H}_{os}|1\rangle & \langle 4|\mathcal{H}_{os}|2\rangle & \langle 4|\mathcal{H}_{os}|3\rangle & \langle 4|\mathcal{H}_{os}|4\rangle
\end{bmatrix}
$$

$$
= \begin{bmatrix}
\delta E_{hy} - \zeta & 0 & 0 & 0 \\
0 & \delta E_{hy} + \zeta & 0 & 0 \\
0 & 0 & \delta E_{hy} + \zeta & 0 \\
0 & 0 & 0 & \delta E_{hy} - \zeta
\end{bmatrix},
$$

(7.56)

where

$$\delta E_{hy} = -2a_1\left(1 - \frac{c_{12}}{c_{11}}\right)\varepsilon,$$

$$\zeta = -b_1\left(1 + \frac{2c_{12}}{c_{11}}\right)\varepsilon,$$

(7.57)

and (7.35) were used. Equations (7.56) and (7.57) represent the energy shifts at $\mathbf{k} = 0$, which indicates that degeneracy of the valence bands is removed.

The orbit-strain interaction Hamiltonian for the conduction band $\mathcal{H}_{sc}$ is given by

$$\mathcal{H}_{sc} = C_1(\varepsilon_{xx} + \varepsilon_{yy} + \varepsilon_{zz}),$$

(7.58)

where C_1 is the deformation potential of the conduction band. The energies of the heavy hole band and the light hole band are given by the eigenvalues of

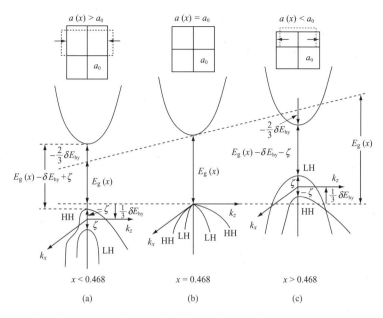

Fig. 7.8. Band structures for the bulk $In_{1-x}Ga_xAs$ layers grown on an InGaAsP layer, which is lattice-matched to an InP substrate: (a) compressive, (b) lattice matching, and (c) tensile

$\mathcal{H}_{LK} + \mathcal{H}_{sv}$ with the help of (7.36) and (7.56). The energy of the conduction band is obtained as the eigenvalue of $\mathcal{H}_{LK} + \mathcal{H}_{sc}$ by using (7.36) and (7.58).

Figure 7.8 shows the band structures for the strained bulk $In_{1-x}Ga_xAs$ layers grown on an InGaAsP layer, which is lattice-matched to an InP substrate. Here, $C_1 = -2a_1$ is assumed, and the dislocations are neglected. From Fig. 7.8, it is found that the compressive strains lower the energy of the heavy hole more than that of the light hole, and the tensile strains lower the energy of the light hole more than that of the heavy hole. It should be noted that the energy of the hole is low with an increase in the height, because the vertical axis represents the energy of the electron.

(c) Strained QWs

In the strained QWs, the Hamiltonian for the valence bands is given by

$$\mathcal{H}_{LK} + \mathcal{H}_{sv} + V(z), \tag{7.59}$$

where $V(z)$ represents a potential of the QWs, which is expressed as

$$V(z) = \begin{cases} 0 & \text{(well)}, \\ \Delta E_v - \delta E_{hy} & \text{(barrier)}. \end{cases} \tag{7.60}$$

If the quantization axis is the z-axis, Schrödinger equation can be solved by replacing $k_z \to -i\,\partial/\partial z$, under the effective mass approximations. The obtained results for the energy bands and the optical gains for the heavy hole (HH) and light hole (LH) are illustrated in Fig. 7.9. Here, the oscillation wavelength is 1.3 μm, and the InGaAsP well is 10 nm thick for a 1.9% tensile strain, lattice matching, and a 1.4% compressive strain. As found in Fig. 7.9, the tensile strains enhance the optical gain for the TM mode, and the compressive strains increase the optical gain for the TE mode.

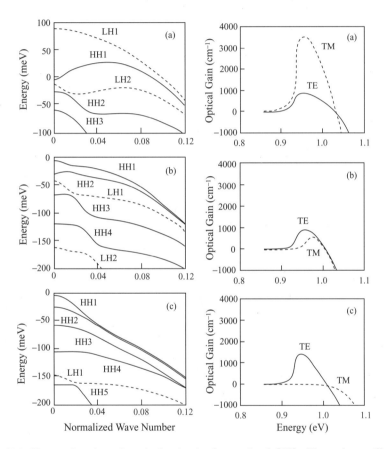

Fig. 7.9. Energy bands and optical gains in the strained QWs. Here, the oscillation wavelength is 1.3 μm, and the InGaAsP well is 10 nm thick for (a) a 1.9% tensile strain, (b) lattice matching, and (c) a 1.4% compressive strain

7.4 Requirements for Fabrication

The QW-LDs have been in practical use. However, the quantum wire LDs and the quantum box LDs, whose quantum effects are more significant than those of the QW-LDs, are still under research because these quantum structures require highly sophisticated fabrication technologies, which can produce

1. highly uniform,
2. highly integrated,
3. various shaped,
4. nanostructures whose quantum effects are significant,
5. with low damage,
6. in a short time.

Fluctuations in size obscure the density of states, and the quantum effects are degraded. Therefore, uniform quantum structures are needed. When we use quantum wires or quantum boxes in the active layers, the optical gain length is short, because of their tiny size. However, as shown in (5.11), to obtain a low threshold gain, we need long optical gain length L, which is only achieved by highly integrated quantum structures. The fourth requirement is indispensable to quantum structures; the others are common to all devices.

To satisfy these requirements, selective epitaxial growth and self-organized epitaxial growth have attracted a lot of interest.

8 Control of Spontaneous Emission

8.1 Introduction

High energy states are unstable. As a result, the electrons in high energy states transit to low energy states in a certain lifetime. The radiation associated with this transition is referred to as spontaneous emission.

Figure 8.1 schematically shows the spontaneous emission in semiconductors, which is the radiative recombination of the electrons in the conduction band and the holes in the valence band. It should be noted that spontaneous emission takes place regardless of incident light, whereas stimulated emission and absorption are induced by incident light.

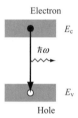

Fig. 8.1. Spontaneous emission in semiconductors

According to quantum mechanics, spontaneous emission takes place due to interactions of atomic systems and the vacuum field. Therefore, the spontaneous emission is considered to be the stimulated emission induced by fluctuations in the vacuum field (zero-point vibrations). It should be noted that the term *stimulated emission* usually represents the stimulated emission induced by incident light.

As described earlier, spontaneous emission is caused by the interactions of atomic systems and the vacuum field. As a result, the mode and the lifetime of the spontaneous emission can be modified by controlling the vacuum field with the optical cavities, which is referred to as *cavity quantum electrodynamics* (QED).

8.2 Spontaneous Emission

8.2.1 Fermi's Golden Rule

With regard to the spontaneous emission rate, we explain *Fermi's golden rule*. As shown in Fig. 8.2, we consider a two-level system, which consists of a *ground state* $|\text{gr}\rangle$ with an energy E_{gr} and an *excited state* $|\text{ex}\rangle$ with an energy E_{ex}. Here, we assume that a resonance spectral linewidth of the optical cavity in units of an angular frequency $\Delta\omega_{\text{c}}$ is much larger than that of the atomic system $\Delta\omega_{\text{a}}$. When the spectrum is Lorentzian, the spectral linewidth is inversely proportional to the lifetime. Hence, the assumption $\Delta\omega_{\text{c}} \gg \Delta\omega_{\text{a}}$ indicates $\tau_{\text{c}} \ll \tau_{\text{a}}$, where $\tau_{\text{c}} = \tau_{\text{ph}}$ is the photon lifetime and τ_{a} is the lifetime of the atomic transition. Therefore, during the transitions of the atoms from the excited state to the ground state, lights are readily emitted outward from the optical cavity. Hence, this transition is irreversible, and absorptions do not take place.

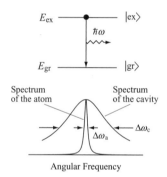

Fig. 8.2. Fermi's golden rule

From a perturbation theory in quantum mechanics, a transition rate W of an atom is given by

$$W = \frac{2\pi}{\hbar} \sum_l \mu_l{}^2 \left(\frac{\hbar\omega_l}{2V\varepsilon} \right) \sin^2(\boldsymbol{k}_l \cdot \boldsymbol{r})(n_l + 1)F(E_{\text{ex}} - E_{\text{gr}} - \hbar\omega_l), \qquad (8.1)$$

where $\hbar = h/2\pi$ is Dirac's constant and h is Planck's constant; μ_l is the electric dipole moment for the lth cavity mode; ω_l is the angular frequency of the lth cavity mode; V is the volume of the optical cavity; ε is the dielectric constant of a material in the optical cavity; $\boldsymbol{k}_l$ is the wave vector of the lth cavity mode; n_l is the photon density; and $F(E_{\text{ex}} - E_{\text{gr}} - \hbar\omega_l)$ is a normalized spectral function.

From (8.1), it is found that $W > 0$ is obtained even when the photons do not exist in the optical cavity ($n_l = 0$). A transition rate for $n_l = 0$ in (8.1) is the spontaneous emission rate W_{sp}, which is expressed as

$$W_{\text{sp}} = \frac{2\pi}{\hbar} \sum_l \mu_l^2 \left(\frac{\hbar \omega_l}{2V\varepsilon} \right) \sin^2(\mathbf{k}_l \cdot \mathbf{r}) F(E_{\text{ex}} - E_{\text{gr}} - \hbar \omega_l). \qquad (8.2)$$

8.2.2 Spontaneous Emission in a Free Space

A *free space* can be regarded as an optical cavity whose size is much larger than a wavelength of a light. Therefore, a spectral distribution of the resonance modes in the free space is (quasi-)continuous. In this case, a sum with respect to a resonance mode l in (8.2) is replaced by integration with respect to $\mathbf{k}$. Hence, the spontaneous emission rate in the free space $W_{\text{sp, free}}$ is written as

$$W_{\text{sp,free}} = \frac{\omega_0^3 n_{\text{rc}}^3}{3\pi c^3 \hbar \varepsilon} \bar{\mu}^2, \qquad (8.3)$$

where n_{rc} is the refractive index of a material in the optical cavity, $\bar{\mu}^2$ is the squared electric dipole moment averaged over all directions of polarizations, $\omega_0 = (E_{\text{ex}} - E_{\text{gr}})/\hbar$ is the angular frequency of a light, and $\sin^2(\mathbf{k}_l \cdot \mathbf{r})$ is replaced by the averaged value $1/2$.

In this case, a transition rate for each mode is small, and it is inversely proportional to the volume of the optical cavity V. The number of modes N contributing to the optical transitions is proportional to V, and the value of N for usual waveguide-type LDs is about 10^5. As a result, the spontaneous emission rate in the free space $W_{\text{sp,free}}$ is independent of V. Moreover, $W_{\text{sp,free}}$ is not affected by the positions of the atoms. Therefore, the spontaneous emission in the free space seems to be governed by the properties of the atoms themselves.

8.2.3 Spontaneous Emission in a Microcavity

The *microcavity* [56] is the optical cavity whose size is on the order of a light wavelength. Hence, the resonance modes in the microcavity are discrete. From (8.2), when the spontaneous emission coupling factor is $\beta_{\text{sp}} = 1$, the spontaneous emission rate in the microcavity $W_{\text{sp,micro}}$ is given by

$$W_{\text{sp,micro}} = \frac{\pi}{\hbar} \mu_l^2 \left(\frac{\hbar \omega_l}{V\varepsilon} \right) \left(\frac{1}{\hbar \Delta \omega_c} \right) \sin^2(\mathbf{k}_l \cdot \mathbf{r}), \qquad (8.4)$$

where $\Delta \omega_c$ is the spectral linewidth of the resonance mode of the optical cavity in units of an angular frequency.

From (8.3) and (8.4), when the angular frequency of the light ω_0 is equal to that of the lth optical cavity mode ω_l ($\omega_0 = \omega_l$), the spontaneous emission rate in the microcavity $W_{\text{sp,micro}}$ and the spontaneous emission rate in the free space $W_{\text{sp,free}}$ are related as

$$\frac{W_{\text{sp,micro}}}{W_{\text{sp,free}}} = \frac{3}{8\pi} \frac{\lambda_l^{\,3}}{n_{\text{rc}}^{\,3}} \frac{1}{V} \left(\frac{\omega_l}{\Delta\omega_c}\right) \sin^2(\boldsymbol{k}_l \cdot \boldsymbol{r}), \qquad (8.5)$$

where λ_l is the resonance wavelength in a vacuum for the lth optical cavity mode.

From (8.5), it is found that enhancement of the spontaneous emission ($W_{\text{sp,micro}} > W_{\text{sp,free}}$) happens when the following three conditions are satisfied: (1) The optical cavity is the microcavity whose size is on the order of a light wavelength ($V \approx \lambda_l^{\,3}/n_{\text{rc}}^{\,3}$), (2) the Q-value of the optical cavity is extremely large ($\omega_l/\Delta\omega_c \gg 1$), and (3) excited atoms are placed at antinodes of the standing wave in the optical cavity where $\sin^2(\boldsymbol{k}_l \cdot \boldsymbol{r}) = 1$.

In contrast, the spontaneous emission is suppressed when the angular frequency of the light ω_0 and that of the lth optical cavity mode ω_l are different from each other ($\omega_0 \neq \omega_l$). If the excited atoms exist at nodes of the standing wave where $\sin^2(\boldsymbol{k}_l \cdot \boldsymbol{r}) = 0$, spontaneous emission is prohibited because $W_{\text{sp,micro}} = 0$.

8.2.4 Fluctuations in the Vacuum Field

In a system consisting of the bosons, such as photons and phonons, the fluctuations in the vacuum field are generated due to Heisenberg's uncertainty principle, even when the bosons do not exist ($n_l = 0$). As a result, there exists a zero-point energy $\hbar\omega_l/2$ in a resonance mode. The spontaneous emission is induced by these fluctuations in the vacuum field, and a theory of Cohen-Tannoudji and others [57] will be briefly explained in the following.

When atoms in the excited state interact with fluctuations in the vacuum field, optical transitions take place from the excited state to the ground state. The state of an atom (electron) $|\Psi\rangle$ is given by a linear combination of the two eigenstates $|\text{ex}\rangle$ and $|\text{gr}\rangle$, which is written as

$$|\Psi\rangle = C_{\text{ex}}|\text{ex}\rangle \exp\left(-\mathrm{i}\frac{E_{\text{ex}}}{\hbar}t\right) + C_g|\text{gr}\rangle \exp\left(-\mathrm{i}\frac{E_{\text{gr}}}{\hbar}t\right). \qquad (8.6)$$

The expectation value of the electric dipole moment is expressed as

$$\langle\Psi|e\boldsymbol{r}|\Psi\rangle = C_{\text{ex}}{}^* C_g \langle\text{ex}|e\boldsymbol{r}|\text{gr}\rangle \exp(\mathrm{i}\omega_0 t) + \text{h.c.}, \qquad (8.7)$$

where e is the elementary charge and h.c. shows the Hermite conjugate.

The electric dipole moment, which vibrates with the angular frequency ω_0, emits an electromagnetic wave with ω_0. Also, the emitted electromagnetic wave reacts on the electric dipole moment. When the atom is in the excited state, the fluctuations in the vacuum field and the electromagnetic wave interact with the same phase, which induces spontaneous emission. While the atom is in the ground state, the interaction takes place with the antiphase and *spontaneous absorption* never happens.

8.3 Microcavity LDs

From (2.27), a low mode density and a large spontaneous emission rate result in a large stimulated emission rate. Hence, a low threshold current is expected in the microcavity LDs. As shown in (8.5), to enhance spontaneous emission, we need microcavities with high Q-values. From this viewpoint, the vertical cavity surface emitting LDs (VCSELs) explained in Section 6.3 are suitable for microcavity LDs.

From (5.20) and (5.21), the rate equations for the carrier concentration n and the photon density S are expressed as

$$\frac{dn}{dt} = \frac{J}{ed} - G(n)S - \frac{n}{\tau_n}, \tag{8.8}$$

$$\frac{dS}{dt} = G(n)S - \frac{S}{\tau_{ph}} + \beta_{sp}\frac{n}{\tau_n}, \tag{8.9}$$

where J is the injection current density, e is the elementary charge, d is the active layer thickness, $G(n)$ is the amplification rate due to the stimulated emission, τ_n is the carrier lifetime, τ_{ph} is the photon lifetime, and β_{sp} is the spontaneous emission coupling factor. It should be noted that $G(n)$ includes an effect of enhancement or suppression of the spontaneous emission caused by the microcavities, and $\tau_r = \tau_n$ is assumed.

When the mode distribution is continuous, from (5.32), the spontaneous emission coupling factor β_{sp} for a wavelength in a vacuum λ_0 is given by

$$\beta_{sp} = \frac{\Gamma_a}{4\pi^2 n_{rc}{}^3 V}\frac{\lambda_0{}^4}{\Delta\lambda}, \tag{8.10}$$

where Γ_a is the optical confinement factor of the active layer and $\Delta\lambda$ is the FWHM of the light emission spectrum in a wavelength region. From (8.10), a small mode volume V and a narrow FWHM $\Delta\lambda$ lead to a large β_{sp}. In microcavities, the mode distribution is discrete, and β_{sp} has a different form from (8.10), but it is no change that β_{sp} increases with a decrease in V.

Figure 8.3 shows the carrier concentration n and the photon density S in the microcavity LDs as a function of the injection current density J. With an increase in β_{sp}, the threshold current density J_{th} decreases and becomes indistinct.

In microcavity LDs, due to enhancement of spontaneous emission, the carrier lifetime τ_n decreases and the stimulated emission rate increases, which enhances $\partial G/\partial n$. As shown in (5.103) and (5.109), the resonance frequency f_r is given by

$$f_r = \frac{1}{2\pi}\sqrt{\frac{1}{\tau_n \tau_{ph}}\frac{J - J_{th}}{J_{th} - J_0'}} = \frac{1}{2\pi}\sqrt{\frac{\partial G}{\partial n}\frac{S_0}{\tau_{ph}}}. \tag{8.11}$$

Therefore, a large f_r, which leads to high-speed modulation, is obtained in microcavity LDs. Note that in the optical cavities with high Q-values, the

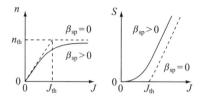

Fig. 8.3. Characteristics of microcavity LDs

photon lifetime τ_{ph} is large, which reduces f_{r}. Hence, we have to design microcavity LDs according to the required modulation speed in application systems.

From the preceding explanations, someone might think that the microcavity LDs have only benefits. However, they have serious problems in electrical resistance and light output. Because the size of microcavity LDs is on the order of a wavelength of a light, their electrical resistances are extremely high. Therefore, with an increase in injection current, large Joule's heat is generated, which suppresses CW operations at room temperature. Even though the threshold current is low, due to large electrical resistance, the power consumptions are not always low, and sometimes they are large. Also, due to small light emission regions, the light outputs are low.

8.4 Photon Recycling

When electric current is injected into semiconductor lasers, spontaneous emission takes place. A fraction of this spontaneous emission is used as the seed of a laser light, and this seed is repeatedly amplified in the optical cavities by the stimulated emission. When the optical gains exceed the optical losses in the optical cavities, laser oscillations start.

In contrast, spontaneous emission other than the seed of the laser light is readily emitted outward from the optical cavities. Therefore, the injected carriers consumed for such spontaneous emission does not contribute to laser oscillations. In *photon recycling*, this wasted spontaneous emission is absorbed, and the carriers are generated [58–61]. Using photon recycling, the carrier concentration in the active layer is larger than that without photon recycling at the same current level, which results in a low threshold current. This concept of photon recycling resembles the *confinement of resonant radiation* in gas lasers.

Figures 8.4 (a) and (b) schematically show the light emission spectra and the optical gain spectra of semiconductor lasers. A peak wavelength in the optical gain spectra is longer than that in the spontaneous emission spectra. Therefore, the active layer can absorb the spontaneous emission located in a shorter wavelength region, which is indicated by slanted lines. Consequently, to achieve efficient photon recycling, we should confine the spontaneously emitted lights to the optical cavities. To avoid wasting the injected carriers, we should also reduce the nonradiative recombination rates and suppress the Joule heating in the active layers so that the carriers may not overflow to the cladding layers.

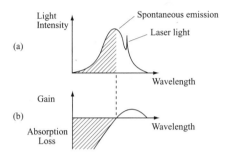

Fig. 8.4. (a) Light emission spectrum and (b) optical gain spectrum

From Fig. 8.4, it is found that a large absorption region in a wavelength leads to efficient photon recycling. As shown in Fig. 8.5, with a decrease in the threshold gain g_{th}, an absorption wavelength region increases. Therefore, semiconductor lasers with low threshold gain as in Fig. 8.5 (b) are suitable for efficient photon recycling.

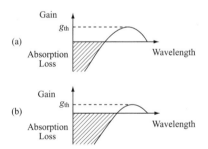

Fig. 8.5. Threshold gain and photon recycling

Figure 8.6 schematically shows the structure of a vertical cavity surface emitting LD, which uses photon recycling. To efficiently confine the spontaneous emission in the optical cavity, the sidewalls of the optical cavity are covered with highly reflective materials.

In photon recycling, limitations in the size of the optical cavities do not exist, as opposed to in microcavity LDs. Consequently, when we use photon recycling, we can avoid an increase in electrical resistance and a decrease in the light output, which were shown in microcavity LDs.

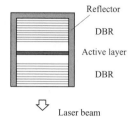

Fig. 8.6. Vertical cavity surface emitting LD with photon recycling

A Cyclotron Resonance

A.1 Fundamental Equations

Let us consider the motion of an electron in a magnetic field. As shown in Fig. A.1, we assume that a uniform magnetic field is applied along the z-axis, and a magnetic flux density is $\boldsymbol{B}(|\boldsymbol{B}| = B_z)$. The *Lorentz force* acting on the electron in this magnetic field has only x- and y-components and does not have a z-component.

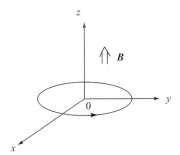

Fig. A.1. Cyclotron motion

When the relaxation term is neglected, an equation of motion for the electron in the magnetic field is given by

$$m \frac{\mathrm{d}\boldsymbol{v}}{\mathrm{d}t} = -e\,(\boldsymbol{v} \times \boldsymbol{B}), \tag{A.1}$$

where m, v, and e are the electron mass, velocity, and elementary charge, respectively. From (A.1), we obtain

$$\ddot{x} = -\omega_{\mathrm{c}} \dot{y}, \quad \ddot{y} = \omega_{\mathrm{c}} \dot{x}, \tag{A.2}$$

where a double dot above x and y shows a second derivative with respect to time, and

$$\omega_{\mathrm{c}} = \frac{eB_z}{m} \tag{A.3}$$

is the *cyclotron angular frequency*.

Assuming $x = 0$, $y = -v_0/\omega_c$, $\dot{x} = v_0$, and $\dot{y} = 0$ at $t = 0$ as an initial condition, and then integrating (A.2) with respect to time result in

$$\dot{x} = -\omega_c y, \quad \dot{y} = \omega_c x, \tag{A.4}$$

where a dot above x and y shows a first derivative with respect to time. Substituting (A.4) into (A.2), we have

$$\ddot{x} = -\omega_c{}^2 x, \quad \ddot{y} = -\omega_c{}^2 y. \tag{A.5}$$

As a result, the solutions are obtained as

$$x = \frac{v_0}{\omega_c} \sin \omega_c t, \quad y = -\frac{v_0}{\omega_c} \cos \omega_c t. \tag{A.6}$$

These solutions show the *cyclotron motion* that is a circular motion with a radius $R = v_0/\omega_c$ about the origin.

When an electromagnetic wave whose electric field is in the xy-plane and the angular frequency ω_c is incident on the semiconductors, the electromagnetic wave is resonantly absorbed by the electrons with cyclotron motions. This phenomenon is called the *cyclotron resonance*.

We assume that the effective mass of the carrier is m^*, the velocity of the carrier is v, the charge of the carrier is q ($q = -e < 0$ for the electron and $q = +e > 0$ for the hole), the electric field is $\boldsymbol{E}$, the magnetic flux density is $\boldsymbol{B}$, and the relaxation time is τ. Then the equation of motion for the carrier is written as

$$m^* \frac{d\boldsymbol{v}}{dt} = q\left(\boldsymbol{E} + \boldsymbol{v} \times \boldsymbol{B}\right) - \frac{m^* \boldsymbol{v}}{\tau}. \tag{A.7}$$

Using the carrier concentration n, we can express the x- and y-components of the current density as $J_i = nqv_i$ ($i = x, y$). From the Joule losses formula, the absorbed energy density per unit time P is obtained as

$$P = \overline{\mathrm{Re}(\boldsymbol{J}) \cdot \mathrm{Re}(\boldsymbol{E})} = \frac{1}{2}\mathrm{Re}(\boldsymbol{J} \cdot \boldsymbol{E}^*), \tag{A.8}$$

where Re indicates a real part. In the following, we consider a right-handed circularly polarized wave, a left-handed circularly polarized wave, and a linearly polarized wave. These electromagnetic waves are assumed to propagate along the z-axis and to have the angular frequency ω.

A.2 Right-Handed Circularly Polarized Wave

The *right-handed circularly polarized wave* is expressed as

$$E_x = E_0 \exp(\mathrm{i}\omega t), \quad E_y = \mathrm{i}E_x = \mathrm{i}E_0 \exp(\mathrm{i}\omega t). \tag{A.9}$$

Substituting (A.9) into (A.7) and using (A.8) results in

$$P = \frac{\sigma_0 E_0{}^2}{(\omega - \omega_c)^2 \tau^2 + 1}, \tag{A.10}$$

where

$$\omega_c = \frac{qB_z}{m^*}, \quad \sigma_0 = \frac{nq^2\tau}{m^*}. \tag{A.11}$$

A.3 Left-Handed Circularly Polarized Wave

The *left-handed circularly polarized wave* is written as

$$E_x = E_0 \exp(\mathrm{i}\omega t), \quad E_y = -\mathrm{i}\,E_x = -\mathrm{i}\,E_0 \exp(\mathrm{i}\omega t). \tag{A.12}$$

Substituting (A.12) into (A.7) and using (A.8) leads to

$$P = \frac{\sigma_0 E_0{}^2}{(\omega + \omega_c)^2 \tau^2 + 1}. \tag{A.13}$$

A.4 Linearly Polarized Wave

When the *linearly polarized wave* has an electric field with only an x-component, which is given by

$$E_x = E_0 \exp(\mathrm{i}\omega t), \tag{A.14}$$

the absorbed energy density per unit time P is represented as

$$P = \frac{1}{4}\sigma_0 E_0{}^2 \left[\frac{1}{(\omega - \omega_c)^2 \tau^2 + 1} + \frac{1}{(\omega + \omega_c)^2 \tau^2 + 1} \right]. \tag{A.15}$$

Here, the first and second terms correspond to the right-handed circularly polarized wave and the left-handed circularly polarized wave, respectively.

A.5 Relationship between Polarization of a Wave and an Effective Mass

From (A.11), it is found that we can obtain the effective mass of the carrier by measuring the cyclotron angular frequency ω_c. The right-handed circularly polarized wave is resonantly absorbed by the carrier, which has a positive effective mass with a positive ω_c, such as the electron in the conduction band in the vicinity of the band edge. The left-handed circularly polarized wave is resonantly absorbed by the carrier, which has a negative effective mass with a negative ω_c, such as the hole in the valence bands in the vicinity of the band edge.

Finally, we show an example of a value for the cyclotron angular frequency ω_c. When the observed effective mass in a magnetic field with the magnetic flux density of $1\,G = 10^{-4}\,T$ is $0.1m$, where m is the electron mass in a vacuum, ω_c is $1.76 \times 10^8\,rad/s$, and the cyclotron frequency $f_c = \omega_c/2\pi$ is $2.80 \times 10^7\,Hz$.

B Time-Independent Perturbation Theory

We consider the following Schrödinger equation expressed as

$$\mathcal{H}\psi = W\psi, \quad \mathcal{H} = \mathcal{H}_0 + \mathcal{H}', \quad \mathcal{H}_0 u_k = E_k u_k, \tag{B.1}$$

where $\mathcal{H}_0$ is an *unperturbed Hamiltonian* whose solutions are already obtained; $\mathcal{H}'$ is a *perturbation* to $\mathcal{H}_0$; ψ and W are an eigenfunction (wave function) and an energy eigenvalue for a perturbed steady state, respectively; and u_k and E_k are an orthonormalized eigenfunction and an energy eigenvalue for the unperturbed Hamiltonian $\mathcal{H}_0$, respectively.

B.1 Nondegenerate Case

The perturbed eigenfunction and the perturbed energy eigenvalue are assumed to be expanded in power series of $\mathcal{H}'$. Here, we replace $\mathcal{H}'$ by $\lambda\mathcal{H}'$ where λ is a parameter, and we express ψ and W with the power series of λ. In obtaining the final results, we set $\lambda = 1$. Under this assumption, ψ and W are written as

$$\begin{aligned}
\psi &= \psi_0 + \lambda\psi_1 + \lambda^2\psi_2 + \lambda^3\psi_3 + \cdots, \\
W &= W_0 + \lambda W_1 + \lambda^2 W_2 + \lambda^3 W_3 + \cdots.
\end{aligned} \tag{B.2}$$

Substituting (B.2) into (B.1) and replacing $\mathcal{H}'$ by $\lambda\mathcal{H}'$, we have

$$\begin{aligned}
(\mathcal{H}_0 + &\lambda\mathcal{H}')(\psi_0 + \lambda\psi_1 + \lambda^2\psi_2 + \lambda^3\psi_3 + \cdots) \\
&= (W_0 + \lambda W_1 + \lambda^2 W_2 + \lambda^3 W_3 + \cdots)(\psi_0 + \lambda\psi_1 + \lambda^2\psi_2 + \lambda^3\psi_3 + \cdots).
\end{aligned} \tag{B.3}$$

If we assume that (B.3) is satisfied for any value of λ, the terms with the common power series of λ on both sides have to be the same, which is expressed as

$$\begin{aligned}
\lambda^0 &: \quad (\mathcal{H}_0 - W_0)\psi_0 = 0, \\
\lambda^1 &: \quad (\mathcal{H}_0 - W_0)\psi_1 = (W_1 - \mathcal{H}')\psi_0, \\
\lambda^2 &: \quad (\mathcal{H}_0 - W_0)\psi_2 = (W_1 - \mathcal{H}')\psi_1 + W_2\psi_0, \\
\lambda^3 &: \quad (\mathcal{H}_0 - W_0)\psi_3 = (W_1 - \mathcal{H}')\psi_2 + W_2\psi_1 + W_3\psi_0, \\
&\qquad\qquad\vdots
\end{aligned} \tag{B.4}$$

From (B.1) and the first equation in (B.4), it is found that ψ_0 is one of the unperturbed eigenfunctions u_ks. Therefore, in the nondegenerate case, we put

$$\psi_0 = u_m, \quad W_0 = E_m. \tag{B.5}$$

As an eigenfunction $\psi_s (s > 0)$ in a perturbed steady state, we consider a solution satisfying the dot product given by

$$(\psi_0, \psi_s) = \int \psi_0^* \psi_s \, \mathrm{d}^3 \boldsymbol{r} = 0. \tag{B.6}$$

Under this condition, multiplying both sides of (B.4) by ψ_0^* from the left and then integrating with respect to all the space results in

$$W_s = \frac{(\psi_0, \mathcal{H}' \psi_{s-1})}{(\psi_0, \psi_0)} = (u_m, \mathcal{H}' \psi_{s-1}), \tag{B.7}$$

where we have used that u_ks are orthonormal functions. From (B.5) and (B.7), we have

$$W_1 = (u_m, \mathcal{H}' \psi_0) = (u_m, \mathcal{H}' u_m) = \langle m | \mathcal{H}' | m \rangle, \tag{B.8}$$

from which it is revealed that the first-order perturbation energy W_1 is the expectation value of $\mathcal{H}'$ for the unperturbed state $|m\rangle$.

Here, by expanding with respect to u_n, we write the first-order perturbed eigenfunction ψ_1 as

$$\psi_1 = \sum_n a_n^{(1)} u_n, \tag{B.9}$$

where $a_n^{(1)}$ is the expansion coefficient. If we calculate $a_n^{(1)}$, an eigenfunction in the first-order perturbation is obtained. First, let us consider a dot product of u_m and (B.9). Because u_n is an orthonormal function, we have the following relation

$$\langle m | n \rangle = \delta_{mn}, \tag{B.10}$$

where δ_{mn} is the Kronecker delta. Using (B.6), we obtain

$$(u_m, \psi_1) = \langle m | \sum_n a_n^{(1)} | n \rangle = \sum_n a_n^{(1)} \langle m | n \rangle$$

$$= a_m^{(1)} = (\psi_0, \psi_1) = 0. \tag{B.11}$$

Substituting (B.9) into the second equation in (B.4) results in

$$\sum_n a_n^{(1)} (\mathcal{H}_0 - E_m) u_n = (W_1 - \mathcal{H}') u_m, \tag{B.12}$$

where (B.5) was used. Taking a dot product of u_k and (B.12) with the help of (B.1) leads to

$$a_k^{(1)}(E_k - E_m) = -\langle k|\mathcal{H}'|m\rangle,$$

$$\therefore \ a_k^{(1)} = \frac{\langle k|\mathcal{H}'|m\rangle}{E_m - E_k} \quad (k \neq m). \tag{B.13}$$

As a result, ψ_1 is written as

$$\psi_1 = \sum_n \frac{\langle n|\mathcal{H}'|m\rangle}{E_m - E_n} u_n. \tag{B.14}$$

Therefore, to the first-order perturbation, the eigenfunction ψ is given by

$$\psi = \psi_0 + \psi_1 = u_m + \sum_n \frac{\langle n|\mathcal{H}'|m\rangle}{E_m - E_n} u_n, \tag{B.15}$$

where we have set $\lambda = 1$.

Putting $s = 2$ in (B.7), the second-order perturbation energy is obtained as

$$W_2 = (u_m, \mathcal{H}'\psi_1) = \sum_n \frac{\langle m|\mathcal{H}'|n\rangle\langle n|\mathcal{H}'|m\rangle}{E_m - E_n}, \tag{B.16}$$

where (B.14) was used. From (B.5), (B.8), and (B.16), to the second-order perturbation, the energy eigenvalue is expressed as

$$W = W_0 + W_1 + W_2$$

$$= E_m + \langle m|\mathcal{H}'|m\rangle + \sum_n \frac{\langle m|\mathcal{H}'|n\rangle\langle n|\mathcal{H}'|m\rangle}{E_m - E_n}. \tag{B.17}$$

Finally, let us calculate an eigenfunction in the second-order perturbation. The second-order perturbed eigenfunction ψ_2 is expanded with u_n as

$$\psi_2 = \sum_n a_n^{(2)} u_n, \tag{B.18}$$

where $a_m^{(2)} = 0$. Substituting (B.5), (B.9), and (B.18) into the third equation in (B.4) gives

$$\sum_n a_n^{(2)}(\mathcal{H}_0 - W_0)u_n = \sum_n a_n^{(1)}(W_1 - \mathcal{H}')u_n + W_2 u_m. \tag{B.19}$$

Taking a dot product of $u_k(k \neq m)$ and (B.19) with the help of (B.1) results in

$$a_k^{(2)}(E_k - E_m) = a_k^{(1)}W_1 - a_n^{(1)}\langle k|\mathcal{H}'|n\rangle. \tag{B.20}$$

From (B.8), (B.13), and (B.20), we obtain

$$a_k^{(2)} = \sum_n \frac{\langle k|\mathcal{H}'|n\rangle\langle n|\mathcal{H}'|m\rangle}{(E_m - E_k)(E_m - E_n)} - \frac{\langle k|\mathcal{H}'|m\rangle\langle m|\mathcal{H}'|m\rangle}{(E_m - E_k)^2}. \tag{B.21}$$

As a result, to the second-order perturbation with $\lambda = 1$, the eigenfunction ψ is written as

$$
\begin{aligned}
\psi &= \psi_0 + \psi_1 + \psi_2 \\
&= u_m + \sum_k u_k \left[\frac{\langle k|\mathcal{H}'|m\rangle}{E_m - E_k}\left(1 - \frac{\langle m|\mathcal{H}'|m\rangle}{E_m - E_k}\right) \right. \\
&\quad \left. + \sum_n \frac{\langle k|\mathcal{H}'|n\rangle\langle n|\mathcal{H}'|m\rangle}{(E_m - E_k)(E_m - E_n)} \right].
\end{aligned}
\tag{B.22}
$$

In summary, the eigenfunction and the energy eigenvalue in the non-degenerate case are expressed as

to *the first-order perturbation*

$$\psi = u_m + \sum_n \frac{\langle n|\mathcal{H}'|m\rangle}{E_m - E_n} u_n, \tag{B.23}$$

$$W = W_0 + W_1 = E_m + \langle m|\mathcal{H}'|m\rangle, \tag{B.24}$$

to *the second-order perturbation*

$$
\begin{aligned}
\psi &= u_m + \sum_k u_k \left[\frac{\langle k|\mathcal{H}'|m\rangle}{E_m - E_k}\left(1 - \frac{\langle m|\mathcal{H}'|m\rangle}{E_m - E_k}\right) \right. \\
&\quad \left. + \sum_n \frac{\langle k|\mathcal{H}'|n\rangle\langle n|\mathcal{H}'|m\rangle}{(E_m - E_k)(E_m - E_n)} \right],
\end{aligned}
\tag{B.25}
$$

$$W = E_m + \langle m|\mathcal{H}'|m\rangle + \sum_n \frac{\langle m|\mathcal{H}'|n\rangle\langle n|\mathcal{H}'|m\rangle}{E_m - E_n}. \tag{B.26}$$

B.2 Degenerate Case

In the perturbation theory in the degenerate case, only the zeroth-order eigenfunction ψ_0 is different from that of the perturbation theory in the nondegenerate case; other processes are common.

Here, we assume that the unperturbed eigenfunctions u_l and u_m ($l \neq m$) have the same unperturbed energy W_0. As a result, we can write ψ_0 and W_0 as

$$\psi_0 = a_l u_l + a_m u_m, \quad W_0 = E_l = E_m. \tag{B.27}$$

Substituting (B.27) into the second equation in (B.4) and then taking dot products with u_l and u_m, we have

$$(\langle l|\mathcal{H}'|l\rangle - W_1) a_l + \langle l|\mathcal{H}'|m\rangle a_m = 0,$$

$$\langle m|\mathcal{H}'|l\rangle a_l + (\langle m|\mathcal{H}'|m\rangle - W_1) a_m = 0. \tag{B.28}$$

In order that (B.28) may have solutions other than $a_l = a_m = 0$, the determinant for the coefficients of a_l and a_m must be zero, which is expressed as

$$(\langle l|\mathcal{H}'|l\rangle - W_1)(\langle m|\mathcal{H}'|m\rangle - W_1) - \langle l|\mathcal{H}'|m\rangle\langle m|\mathcal{H}'|l\rangle = 0. \tag{B.29}$$

Solving (B.29) with respect to W_1, we have

$$W_1 = \frac{1}{2}\left(\langle l|\mathcal{H}'|l\rangle + \langle m|\mathcal{H}'|m\rangle\right)$$

$$\pm \frac{1}{2}\left[\left(\langle l|\mathcal{H}'|l\rangle - \langle m|\mathcal{H}'|m\rangle\right)^2 + 4\langle l|\mathcal{H}'|m\rangle\langle m|\mathcal{H}'|l\rangle\right]^{1/2}. \tag{B.30}$$

From (B.30), it is found that degeneracy is removed in the first-order perturbation, so long as the terms in $[\cdots]$ do not vanish. When degeneracy is removed and W_1 has two solutions, a_l and a_m are obtained from (B.28), and they are used to calculate the wave functions. Substituting (B.9) and (B.27) into the second equation in (B.4) and then taking a dot product with $u_k(k \neq l, m)$, we obtain

$$a_k^{(1)}(E_k - E_m) = -\langle k|\mathcal{H}'|m\rangle a_m - \langle k|\mathcal{H}'|l\rangle a_l. \tag{B.31}$$

If we assume that $a_l^{(1)} = a_m^{(1)} = 0$, (B.6) with $s = 1$ is satisfied. Using $a_k^{(1)}$ in (B.31), we can obtain the eigenfunction to the first-order perturbation.

To the second-order perturbation, substituting (B.27) into the third equation in (B.4) and then taking dot products with u_l and u_m, we have

$$\sum_n \langle l|\mathcal{H}'|n\rangle a_n^{(1)} - W_2\, a_l = 0,$$

$$\sum_n \langle m|\mathcal{H}'|n\rangle a_n^{(1)} - W_2\, a_m = 0. \tag{B.32}$$

Substituting (B.31) into (B.32) leads to

$$\left(\sum_n \frac{\langle l|\mathcal{H}'|n\rangle\langle n|\mathcal{H}'|l\rangle}{E_m - E_n} - W_2\right) a_l + \sum_n \frac{\langle l|\mathcal{H}'|n\rangle\langle n|\mathcal{H}'|m\rangle}{E_m - E_n} a_m = 0,$$

$$\sum_n \frac{\langle m|\mathcal{H}'|n\rangle\langle n|\mathcal{H}'|l\rangle}{E_m - E_n} a_l + \left(\sum_n \frac{\langle m|\mathcal{H}'|n\rangle\langle n|\mathcal{H}'|m\rangle}{E_m - E_n} - W_2\right) a_m = 0. \tag{B.33}$$

By solving (B.33), we can obtain the energy eigenvalues to the second-order perturbation.

C Time-Dependent Perturbation Theory

C.1 Fundamental Equation

We consider a *time-dependent Schrödinger equation* such as

$$i\hbar \frac{\partial \psi}{\partial t} = \mathcal{H}\psi, \quad \mathcal{H} = \mathcal{H}_0 + \mathcal{H}'(t), \quad \mathcal{H}_0 u_k = E_k u_k, \tag{C.1}$$

where $\mathcal{H}_0$ is an unperturbed steady-state Hamiltonian whose solutions are already obtained, $\mathcal{H}'(t)$ is a time-dependent perturbation to $\mathcal{H}_0$, ψ is a perturbed eigenfunction (wave function), and u_k and E_k are an orthonormal eigenfunction and an energy eigenvalue for $\mathcal{H}_0$, respectively. When $\mathcal{H}'(t) \neq 0$, transitions take place between the eigenstates of $\mathcal{H}_0$.

Using a solution for an unperturbed wave equation given by

$$u_n\, e^{-i\, E_n t/\hbar}, \tag{C.2}$$

we can expand a solution ψ as

$$\psi = \sum_n a_n(t) u_n\, e^{-i\, E_n t/\hbar}, \tag{C.3}$$

where $a_n(t)$ is a time-dependent expansion coefficient. Substituting (C.3) into (C.1) results in

$$\sum_n i\hbar\, \dot{a}_n u_n\, e^{-i\, E_n t/\hbar} + \sum_n a_n E_n u_n\, e^{-i\, E_n t/\hbar}$$
$$= \sum_n a_n \left[\mathcal{H}_0 + \mathcal{H}'(t) \right] u_n\, e^{-i\, E_n t/\hbar}$$
$$= \sum_n \left(a_n E_n u_n\, e^{-i\, E_n t/\hbar} + a_n(t)\mathcal{H}'(t) u_n\, e^{-i\, E_n t/\hbar} \right), \tag{C.4}$$

where a dot above the expansion coefficient a_n denotes a first derivative with respect to time. From (C.4), we have

$$\sum_n i\hbar\, \dot{a}_n u_n\, e^{-i\, E_n t/\hbar} = \sum_n a_n(t)\mathcal{H}'(t) u_n\, e^{-i\, E_n t/\hbar}. \tag{C.5}$$

Taking a dot product of (C.5) and u_k leads to

$$i\hbar \dot{a}_k e^{-iE_kt/\hbar} = \sum_n \langle k|\mathcal{H}'(t)|n\rangle a_n e^{-iE_nt/\hbar}. \tag{C.6}$$

When we define the *Bohr angular frequency* ω_{kn} as

$$\omega_{kn} \equiv \frac{E_k - E_n}{\hbar}, \tag{C.7}$$

we can rewrite (C.6) as

$$\dot{a}_k = \frac{1}{i\hbar} \sum_n \langle k|\mathcal{H}'(t)|n\rangle a_n e^{i\omega_{kn}t}. \tag{C.8}$$

Replacing $\mathcal{H}'(t)$ in (C.8) by $\lambda\mathcal{H}'(t)$, we expand a_n with λ as

$$a_n = a_n^{(0)} + \lambda a_n^{(1)} + \lambda^2 a_n^{(2)} + \cdots. \tag{C.9}$$

Substituting (C.9) into (C.8) gives

$$\dot{a}_k^{(0)} + \lambda \dot{a}_k^{(1)} + \lambda^2 \dot{a}_k^{(2)} + \cdots$$
$$= \frac{1}{i\hbar} \sum_n \langle k|\lambda\mathcal{H}'(t)|n\rangle (a_n^{(0)} + \lambda a_n^{(1)} + \lambda^2 a_n^{(2)} + \cdots) e^{i\omega_{kn}t}. \tag{C.10}$$

In order that any value of λ may satisfy (C.10), we need

$$\dot{a}_k^{(0)} = 0, \quad \dot{a}_k^{(s+1)} = \frac{1}{i\hbar} \sum_n \langle k|\mathcal{H}'(t)|n\rangle a_n^{(s)} e^{i\omega_{kn}t}. \tag{C.11}$$

From the first equation in (C.11), it is revealed that a_k is independent of time in the zeroth-order perturbation. From the second equation in (C.11), once a lower order $a_k^{(s)}$ is obtained, a next higher order $a_k^{(s+1)}$ can be calculated. Here, we assume

$$a_k^{(0)} = \langle k|m\rangle = \delta_{km}, \tag{C.12}$$

which means that the initial state $|m\rangle$ is a definite unperturbed energy state. Using (C.11) and (C.12), we obtain

$$\dot{a}_k^{(1)} = \frac{1}{i\hbar} \langle k|\mathcal{H}'(t)|m\rangle e^{i\omega_{km}t}. \tag{C.13}$$

Assuming that $a_k^{(1)}$ is zero before perturbation is applied, that is, $a_k^{(1)} = 0$ at $t = -\infty$, integrating (C.13) with respect to time results in

$$a_k^{(1)} = \frac{1}{i\hbar} \int_{-\infty}^{t} \langle k|\mathcal{H}'(t')|m\rangle e^{i\omega_{km}t'} \, dt', \tag{C.14}$$

which is the expansion coefficient for the first-order perturbation.

C.2 Harmonic Perturbation

When a *harmonic perturbation* is applied at $t = 0$ and removed at $t = t_0$, we have

$$\langle k|\mathcal{H}'(t')|m\rangle = 2\langle k|\mathcal{H}'|m\rangle \sin \omega t', \tag{C.15}$$

where $\langle k|\mathcal{H}'|m\rangle$ is independent of time. Substituting (C.15) into (C.14) leads to

$$
\begin{aligned}
a_k^{(1)}(t \geq t_0) &= \frac{2}{i\hbar} \langle k|\mathcal{H}'|m\rangle \int_{-\infty}^{t_0} \sin \omega t' \, e^{i\omega_{km}t'} \, dt' \\
&= -\frac{\langle k|\mathcal{H}'|m\rangle}{i\hbar} \left[\frac{e^{i(\omega_{km}+\omega)t_0} - 1}{\omega_{km} + \omega} - \frac{e^{i(\omega_{km}-\omega)t_0} - 1}{\omega_{km} - \omega} \right].
\end{aligned} \tag{C.16}
$$

On the right-hand side of (C.16), the first term is dominant for $\omega_{km} \approx -\omega$, that is, $E_k \approx E_m - \hbar\omega$, and the second term is dominant for $\omega_{km} \approx \omega$, that is, $E_k \approx E_m + \hbar\omega$. In other words, to the first-order perturbation, due to the harmonic perturbation (C.15), Planck's quantum energy $\hbar\omega$ is transferred to or received from the system.

When the initial state is $|m\rangle$ and the final state is $|k\rangle$, let us calculate the probability that the system is in the final state $|k\rangle$ after the perturbation is removed.

For $E_k < E_m$, the first term in (C.16) is dominant, which results in

$$|a_k^{(1)}(t \geq t_0)|^2 = \frac{4|\langle k|\mathcal{H}'|m\rangle|^2}{\hbar^2(\omega_{km} + \omega)^2} \sin^2\left[\frac{1}{2}(\omega_{km} + \omega)t_0\right]. \tag{C.17}$$

For $E_k > E_m$, the second term in (C.16) is dominant, which leads to

$$|a_k^{(1)}(t \geq t_0)|^2 = \frac{4|\langle k|\mathcal{H}'|m\rangle|^2}{\hbar^2(\omega_{km} - \omega)^2} \sin^2\left[\frac{1}{2}(\omega_{km} - \omega)t_0\right]. \tag{C.18}$$

C.3 Transition Probability

Figure C.1 shows a plot of (C.18). Here, we assume that there is a group of final states $|k\rangle$ whose energy E_k is nearly equal to $E_m + \hbar\omega$, and $\langle k|\mathcal{H}'|m\rangle$ is roughly independent of $|k\rangle$. In this case, the probability of finding the system in one of these states $|k\rangle$ is obtained by integrating (C.18) with respect to ω_{km}. This integration is shown by an area enclosed by the horizontal line and the curve in Fig. C.1. The height of the main peak is proportional to t_0^2, and the spectral linewidth is in proportion to t_0^{-1}. As a result, the probability of finding the system in one of these states $|k\rangle$ is proportional to t_0.

It should be noted that the uncertainty principle is satisfied, because the spectral linewidth, which represents uncertainty in the energy, is in proportion to t_0^{-1}. In addition, because of $E_k \approx E_m + \hbar\omega$, the energy conservation law is also satisfied. It is interesting that the uncertainty principle and the energy conservation law are automatically obtained without particular assumptions.

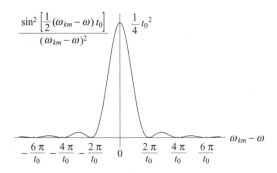

Fig. C.1. Transition probability

The transition probability per unit time w is given by

$$w = \frac{1}{t_0} \int |a_k^{(1)}(t \geq t_0)|^2 \rho(k) \, dE_k, \tag{C.19}$$

where $\rho(k) \, dE_k$ is the number of the final states with the energy between E_k and $E_k + dE_k$. With an increase in t_0, the spectral linewidth decreases, and $\langle k|\mathcal{H}'|m\rangle$ and $\rho(k)$ are considered to be approximately independent of E_k. In this case, $\langle k|\mathcal{H}'|m\rangle$ and $\rho(k)$ can be taken outside the integral in (C.19). Therefore, the transition probability per unit time w is obtained as

$$
\begin{aligned}
w &= \frac{4|\langle k|\mathcal{H}'|m\rangle|^2}{\hbar^2 t_0} \rho(k) \int_{-\infty}^{\infty} \frac{\sin^2\left[\frac{1}{2}(\omega_{km} - \omega)t_0\right]}{(\omega_{km} - \omega)^2} \hbar \, d\omega_{km} \\
&= \frac{2\pi}{\hbar} \rho(k) \, |\langle k|\mathcal{H}'|m\rangle|^2.
\end{aligned}
\tag{C.20}
$$

C.4 Electric Dipole Interaction (Semiclassical Treatment)

A classical Hamiltonian $\mathcal{H}_c$ for a particle, which has a mass m and an electric charge $-e$ in an electromagnetic field, is written as

$$\mathcal{H}_c = \frac{1}{2m}(\boldsymbol{p} + e\boldsymbol{A})^2 - e\phi,$$

$$\boldsymbol{E} = -\frac{\partial \boldsymbol{A}}{\partial t} - \nabla\phi, \tag{C.21}$$

where $\boldsymbol{E}$ is the electric field, $\boldsymbol{A}$ is the *vector potential*, and ϕ is the *scalar potential*.

First, before calculating the transition probability, we will derive (C.21). The *Lagrangian* L for a particle with the mass m and the electric charge $-e$ in an electromagnetic field is given by

$$L = \frac{1}{2}m\dot{\boldsymbol{q}}^2 + e\phi(\boldsymbol{q}) - e\dot{\boldsymbol{q}} \cdot \boldsymbol{A}(\boldsymbol{q}), \tag{C.22}$$

where $\boldsymbol{q}$ is a generalized coordinate. Validity of (C.22) will be shown in the following.

The *Lagrange equation* is expressed as

$$\frac{dt}{dt}\frac{\partial L}{\partial \dot{\boldsymbol{q}}} - \frac{\partial L}{\partial \boldsymbol{q}} = 0. \tag{C.23}$$

As an example, we consider an x-component in the xyz-coordinate system. From (C.22), we have

$$\frac{\partial L}{\partial \dot{x}} = m\dot{x} - eA_x, \tag{C.24}$$

$$\frac{dt}{dt}\frac{\partial L}{\partial \dot{x}} = m\ddot{x} - e\left(\frac{\partial A_x}{\partial t} + \dot{x}\frac{\partial A_x}{\partial x} + \dot{y}\frac{\partial A_x}{\partial y} + \dot{z}\frac{\partial A_x}{\partial z}\right), \tag{C.25}$$

$$\frac{\partial L}{\partial x} = e\frac{\partial \phi}{\partial x} - e\left(\dot{x}\frac{\partial A_x}{\partial x} + \dot{y}\frac{\partial A_y}{\partial x} + \dot{z}\frac{\partial A_z}{\partial x}\right). \tag{C.26}$$

Substituting (C.24)–(C.26) into (C.23) results in

$$m\ddot{x} - e\frac{\partial \phi}{\partial x} - e\left[\frac{\partial A_x}{\partial t} + \dot{y}\left(\frac{\partial A_x}{\partial y} - \frac{\partial A_y}{\partial x}\right) + \dot{z}\left(\frac{\partial A_x}{\partial z} - \frac{\partial A_z}{\partial x}\right)\right] = 0 \tag{C.27}$$

or

$$m\frac{d^2x}{dt^2} = -eE_x - e(\boldsymbol{v} \times \boldsymbol{B})_x,$$

$$E_x = -\frac{\partial \phi}{\partial x} - \frac{\partial A_x}{\partial t}, \quad \boldsymbol{B} = \mathrm{rot}\,\boldsymbol{A}, \tag{C.28}$$

which is known as the *Lorentz equation*, and validity of (C.22) has been proved. Similar results are obtained for the y- and z-components.

Using the Lagrangian L, a momentum $\boldsymbol{p}$ of the particle is defined as

$$\boldsymbol{p} \equiv \frac{\partial L}{\partial \dot{\boldsymbol{q}}} = m\dot{\boldsymbol{q}} - e\boldsymbol{A}. \tag{C.29}$$

As a result, the Hamiltonian $\mathcal{H}$ for the particle is obtained as

$$\mathcal{H}(\boldsymbol{p}, \boldsymbol{q}) \equiv \boldsymbol{p} \cdot \dot{\boldsymbol{q}} - L$$

$$= m\dot{q}^2 - e\,\dot{\boldsymbol{q}} \cdot \boldsymbol{A} - \frac{1}{2} m\dot{q}^2 - e\phi + e\,\dot{\boldsymbol{q}} \cdot \boldsymbol{A}$$

$$= \frac{1}{2} m\dot{q}^2 - e\phi$$

$$= \frac{1}{2m} (\boldsymbol{p} + e\boldsymbol{A})^2 - e\phi, \tag{C.30}$$

which is the same as (C.21).

When $\nabla \cdot \boldsymbol{A} = 0$ and $\phi = 0$, the Schrödinger equation is written as

$$i\hbar \frac{\partial \psi}{\partial t} = [\mathcal{H}_0 + \mathcal{H}'(t)]\psi,$$

$$\mathcal{H}_0 = -\frac{\hbar^2}{2m} \nabla^2 + V(\boldsymbol{r}), \quad \mathcal{H}'(t) = -\frac{i e\hbar}{m} \boldsymbol{A} \cdot \nabla, \tag{C.31}$$

where (C.21) was used. If we express an electric field $\boldsymbol{E}$ as

$$\boldsymbol{E} = \hat{\boldsymbol{x}} \, E_0 \, \exp[i\,(\omega t - kz)], \tag{C.32}$$

where $\hat{\boldsymbol{x}}$ is a unit vector along the x-axis, the vector potential $\boldsymbol{A}$ is written as

$$\boldsymbol{A} = \hat{\boldsymbol{x}} \, \frac{i\,E_0}{\omega} \, \exp[i\,(\omega t - kz)], \tag{C.33}$$

which leads to

$$|\boldsymbol{A}|^2 = |\boldsymbol{A}^* \cdot \boldsymbol{A}| = \frac{E_0{}^2}{\omega^2}. \tag{C.34}$$

To consider the energy flow, we calculate the *Poynting vector* $\boldsymbol{S}$. Using

$$\boldsymbol{B} = \mathrm{rot}\,\boldsymbol{A} = \mu_0 \boldsymbol{H}, \tag{C.35}$$

where $\boldsymbol{B}$ is the magnetic flux density and $\boldsymbol{H}$ is the magnetic field, we have

$$|\boldsymbol{S}| = |\boldsymbol{E} \times \boldsymbol{H}| = \frac{1}{2} E_0{}^2 \frac{k}{\mu_0 \omega} = \frac{1}{2} E_0{}^2 \frac{n_\mathrm{r}}{\mu_0 c} = \frac{1}{2} \varepsilon_0 E_0{}^2 n_\mathrm{r} c. \tag{C.36}$$

Here, n_r is the refractive index of a material. Because (C.36) is equal to the energy flow of $n_\mathrm{ph}\hbar\omega\,c/n_\mathrm{r}$, we obtain

$$E_0{}^2 = \frac{2n_\mathrm{ph}\hbar\omega}{\varepsilon_0\,n_\mathrm{r}{}^2}, \tag{C.37}$$

where n_ph is the photon density. From (C.34) and (C.37), we obtain

$$|\boldsymbol{A}|^2 = \frac{2n_\mathrm{ph}\hbar}{\varepsilon_0 n_\mathrm{r}{}^2\,\omega}. \tag{C.38}$$

Also, (C.31) gives

$$\langle k|\mathcal{H}'|m\rangle = -\frac{\mathrm{i}\,e\hbar}{m}\,\boldsymbol{A}\,\langle k|\nabla|m\rangle = \frac{e}{m}\,\boldsymbol{A}\,\langle k|\boldsymbol{p}|m\rangle$$

$$= e\boldsymbol{A}\,\frac{\mathrm{d}}{\mathrm{d}t}\,\langle k|\boldsymbol{r}|m\rangle = \omega_{km}\,\boldsymbol{A}\,\langle k|e\boldsymbol{r}|m\rangle, \tag{C.39}$$

where the matrix element includes the *electric dipole moment* $e\boldsymbol{r}$. Therefore, the transition due to this perturbation is referred to as the *electric dipole transition*.

When the perturbation is applied at $t = 0$ and removed at $t = t_0$, substituting (C.33) and (C.39) into (C.14) results in

$$a_k^{(1)}(t \geq t_0) = \frac{eE_0}{\hbar\omega m}\,\langle k|\boldsymbol{p}|m\rangle\,\mathrm{e}^{-\mathrm{i}\,kz}\,\frac{\mathrm{e}^{\mathrm{i}\,(\omega_{km}+\omega)t_0} - 1}{\omega_{km} + \omega}. \tag{C.40}$$

When there is a flow of a photon per unit volume, by substituting (C.40) into (C.19), we can obtain the transition probability per unit time w as

$$w = \frac{e^2 h}{2m^2\varepsilon_0 n_r{}^2 E_{km}}|\langle k|\boldsymbol{p}|m\rangle|^2 \qquad (E_{km} \equiv \hbar\omega_{km}), \tag{C.41}$$

where the spectral linewidth of $|\boldsymbol{A}|^2$ is assumed to be narrow.

It should be noted that the transition probability per unit time w in (C.41) is the transition rate for the stimulated emission B in (2.35).

D TE Mode and TM Mode

D.1 Fundamental Equation

Figure D.1 shows a two-dimensional optical waveguide in which the guiding layer is sandwiched between the cladding layer and the substrate. We assume that each layer is uniform and the refractive indexes of the guiding layer, the cladding layer, and the substrate are n_f, n_c, and n_s, respectively, where $n_f > n_s \geq n_c$ is satisfied to confine a light in the guiding layer. In usual optical waveguides, $n_f - n_s$ is 10^{-3}–10^{-1}.

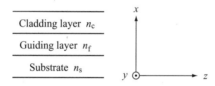

Fig. D.1. Cross-sectional view of a two-dimensional optical waveguide

When the electric current does not flow, *Maxwell's equations* are written as

$$\mathrm{rot}\boldsymbol{E} = \nabla \times \boldsymbol{E} = -\mu_0 \frac{\partial \boldsymbol{H}}{\partial t}, \quad \mathrm{rot}\boldsymbol{H} = \nabla \times \boldsymbol{H} = \varepsilon_0 n_r^2 \frac{\partial \boldsymbol{E}}{\partial t}, \tag{D.1}$$

where $\boldsymbol{E}$ is the electric field, $\boldsymbol{H}$ is the magnetic field, μ_0 is permeability in a vacuum, ε_0 is permittivity in a vacuum, and n_r is the refractive index of a material. We can also express (D.1) with each component as

$$
\begin{aligned}
\frac{\partial E_z}{\partial y} - \frac{\partial E_y}{\partial z} &= -\mu_0 \frac{\partial H_x}{\partial t}, & \frac{\partial H_z}{\partial y} - \frac{\partial H_y}{\partial z} &= \varepsilon_0 n_r^2 \frac{\partial E_x}{\partial t}, \\
\frac{\partial E_x}{\partial z} - \frac{\partial E_z}{\partial x} &= -\mu_0 \frac{\partial H_y}{\partial t}, & \frac{\partial H_x}{\partial z} - \frac{\partial H_z}{\partial x} &= \varepsilon_0 n_r^2 \frac{\partial E_y}{\partial t}, \\
\frac{\partial E_y}{\partial x} - \frac{\partial E_x}{\partial y} &= -\mu_0 \frac{\partial H_z}{\partial t}, & \frac{\partial H_y}{\partial x} - \frac{\partial H_x}{\partial y} &= \varepsilon_0 n_r^2 \frac{\partial E_z}{\partial t}.
\end{aligned}
\tag{D.2}
$$

We assume that a light propagates toward a positive direction on the z-axis with a propagation constant β, and its electric field $\boldsymbol{E}$ and magnetic field H are written as

$$E = E(x, y) \exp[i(\omega t - \beta z)],$$
$$H = H(x, y) \exp[i(\omega t - \beta z)]. \tag{D.3}$$

In a two-dimensional optical waveguide with an abrupt index profile as shown in Fig. D.1, the lightwave along the y-axis is uniform, which results in $\partial/\partial y = 0$. Also, from (D.3), we have $\partial/\partial t = i\omega$ and $\partial/\partial z = -i\beta$. Therefore, (D.2) reduces to

$$i\beta E_y = -i\omega\mu_0 H_x, \quad i\beta H_y = i\varepsilon_0 n_\mathrm{r}^2 \omega E_x,$$
$$-i\beta E_x - \frac{\partial E_z}{\partial x} = -i\omega\mu_0 H_y, \quad -i\beta H_x - \frac{\partial H_z}{\partial x} = i\varepsilon_0 n_\mathrm{r}^2 \omega E_y, \tag{D.4}$$
$$\frac{\partial E_y}{\partial x} = -i\omega\mu_0 H_z, \quad \frac{\partial H_y}{\partial x} = i\varepsilon_0 n_\mathrm{r}^2 \omega E_z.$$

If there are no electric charges, $\nabla \cdot E = 0$ is satisfied. Hence, with the help of the vector analysis formula, we obtain

$$\nabla \times \nabla \times E = \nabla(\nabla \cdot E) - \nabla^2 E = -\nabla^2 E. \tag{D.5}$$

Substituting (D.1) and (D.3) into (D.5) leads to

$$-\nabla^2 E = -\mu_0 \frac{\partial}{\partial t}(\nabla \times H) = -\varepsilon_0 \mu_0 n_\mathrm{r}^2 \frac{\partial^2}{\partial t^2} E = \frac{n_\mathrm{r}^2 \omega^2}{c^2} E, \tag{D.6}$$

where c is the speed of light in a vacuum.

D.2 TE Mode

The electromagnetic wave whose electric field E does not have a component E_z along the propagation direction is called the transverse electric (TE) wave. For a TE wave with $E_z = 0$, (D.4) reduces to

$$i\beta E_y = -i\omega\mu_0 H_x, \quad i\beta H_y = i\varepsilon_0 n_\mathrm{r}^2 \omega E_x,$$
$$-i\beta E_x = -i\omega\mu_0 H_y, \quad -i\beta H_x - \frac{\partial H_z}{\partial x} = i\varepsilon_0 n_\mathrm{r}^2 \omega E_y, \tag{D.7}$$
$$\frac{\partial E_y}{\partial x} = -i\omega\mu_0 H_z, \quad \frac{\partial H_y}{\partial x} = 0.$$

For brevity, if we put $H_y = 0$, (D.7) results in

$$E_x = 0, \quad H_x = -\frac{\beta}{\omega\mu_0} E_y, \quad H_z = -\frac{1}{i\omega\mu_0} \frac{\partial E_y}{\partial x}. \tag{D.8}$$

Substituting the first equation in (D.8) into (D.6) leads to

$$\frac{\partial^2}{\partial x^2} E_y + \left(\frac{n_r^2 \omega^2}{c^2} - \beta^2\right) E_y = \frac{\partial^2}{\partial x^2} E_y + (k_0{}^2 n_r{}^2 - \beta^2) E_y = 0, \qquad (D.9)$$

where $k_0 = \omega/c$.

In summary, the electric field and the magnetic field for the TE mode obey the following equations given by

$$E_x = E_z = 0, \qquad \frac{\partial^2}{\partial x^2} E_y + (k_0{}^2 n_r{}^2 - \beta^2) E_y = 0,$$

$$H_x = -\frac{\beta}{\omega\mu_0} E_y, \quad H_y = 0, \quad H_z = -\frac{1}{i\omega\mu_0} \frac{\partial E_y}{\partial x}. \qquad (D.10)$$

We select a coordinate system, as illustrated in Fig. D.2, and assume that the angle of incidence is θ_f, which is the same as the angle of reflection. If we express the amplitudes of the electric fields in the cladding layer, the guiding layer, and the substrate as E_c, E_f, and E_s, respectively, the electric field E_y is written as

$$\begin{aligned} E_y &= E_c \exp(-\gamma_c x) &&: \text{cladding layer,} \\ E_y &= E_f \cos(k_x x + \phi_c) &&: \text{guiding layer,} \\ E_y &= E_s \exp[\gamma_s(x+h)] &&: \text{substrate,} \end{aligned} \qquad (D.11)$$

where

$$\beta = k_0 N, \quad N = n_f \sin \theta_f, \qquad (D.12)$$

$$\gamma_c = k_0 \sqrt{N^2 - n_c{}^2}, \quad k_x = k_0 \sqrt{n_f{}^2 - N^2}, \quad \gamma_s = k_0 \sqrt{N^2 - n_s{}^2}. \qquad (D.13)$$

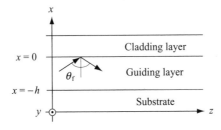

Fig. D.2. Cross-sectional view of a two-dimensional optical waveguide

At each interface, the tangent of the electric field E_y and that of the magnetic field H_z are continuous. Therefore, we obtain

$$E_c = E_f \cos \phi_c, \quad \tan \phi_c = \frac{\gamma_c}{k_x} \qquad : x = 0, \qquad (D.14)$$

$$E_s = E_f \cos(k_x h - \phi_c), \quad \tan(k_x h - \phi_c) = \frac{\gamma_s}{k_x} \qquad : x = -h. \qquad (D.15)$$

From (D.14) and (D.15), the eigenvalue equation is expressed as

$$k_x h = m\pi + \tan^{-1}\frac{\gamma_c}{k_x} + \tan^{-1}\frac{\gamma_s}{k_x}, \tag{D.16}$$

where m is a nonnegative integer. Because the following relation

$$k_x = k_0 n_f \cos\theta_f, \quad \tan^{-1}\frac{\gamma_c}{k_x} = \phi_c, \quad \tan^{-1}\frac{\gamma_s}{k_x} = \phi_s \tag{D.17}$$

is satisfied, (D.16) is the same as (3.16).

D.3 TM Mode

The electromagnetic wave whose magnetic field $\boldsymbol{H}$ does not have a component H_z along the propagation direction is called the transverse magnetic (TM) wave. For the TM wave with $H_z = 0$, (D.4) reduces to

$$i\beta E_y = -i\omega\mu_0 H_x, \quad i\beta H_y = i\varepsilon_0 n_r{}^2 \omega E_x,$$
$$-i\beta E_x - \frac{\partial E_z}{\partial x} = -i\omega\mu_0 H_y, \quad -i\beta H_x = i\varepsilon_0 n_r{}^2 \omega E_y, \tag{D.18}$$
$$\frac{\partial E_y}{\partial x} = 0, \quad \frac{\partial H_y}{\partial x} = i\varepsilon_0 n_r{}^2 \omega E_z.$$

For brevity, if we put $E_y = 0$, (D.18) results in

$$H_x = 0, \quad E_x = \frac{\beta}{\varepsilon_0 n_r{}^2 \omega}H_y, \quad E_z = \frac{1}{i\varepsilon_0 n_r{}^2 \omega}\frac{\partial H_y}{\partial x}, \tag{D.19}$$
$$H_y = \frac{\beta}{\omega\mu_0}E_x + \frac{1}{i\omega\mu_0}\frac{\partial E_z}{\partial x}. \tag{D.20}$$

Substituting (D.19) into (D.20) leads to

$$\frac{\partial^2}{\partial x^2}H_y + (k_0{}^2 n_r{}^2 - \beta^2)H_y = 0. \tag{D.21}$$

In summary, the electric field and the magnetic field for the TM mode obey the following equation given by

$$E_x = \frac{\beta}{\varepsilon_0 n_r{}^2 \omega}H_y, \quad E_y = 0, \quad E_z = \frac{1}{i\varepsilon_0 n_r{}^2 \omega}\frac{\partial H_y}{\partial x},$$
$$\tag{D.22}$$
$$H_x = H_z = 0, \quad \frac{\partial^2}{\partial x^2}H_y + (k_0{}^2 n_r{}^2 - \beta^2)H_y = 0.$$

If we express the amplitudes of the magnetic fields in the cladding layer, the guiding layer, and the substrate as H_c, H_f, and H_s, the magnetic field H_y is written as

$$H_y = H_c \exp(-\gamma_c x) \quad : \text{cladding layer,}$$
$$H_y = H_f \cos(k_x x + \phi_c) \quad : \text{guiding layer,} \qquad \text{(D.23)}$$
$$H_y = H_s \exp[\gamma_s(x + h)] : \text{substrate.}$$

At each interface, the tangent of the magnetic field H_y and that of the electric field E_z are continuous, respectively. Therefore, we obtain

$$H_c = H_f \cos \phi_c, \quad \frac{\gamma_c}{n_c{}^2} H_c = \frac{k_x}{n_f{}^2} H_f \sin \phi_c \quad : x = 0, \qquad \text{(D.24)}$$

$$H_s = H_f \cos(k_x h - \phi_c), \quad \frac{\gamma_s}{n_s{}^2} H_s = \frac{k_x}{n_f{}^2} H_f \sin(k_x h - \phi_c) \quad : x = -h.$$
$$\text{(D.25)}$$

From (D.24) and (D.25), we obtain the following eigenvalue equation

$$k_x h = m\pi + \tan^{-1}\left[\left(\frac{n_f}{n_c}\right)^2 \frac{\gamma_c}{k_x}\right] + \tan^{-1}\left[\left(\frac{n_f}{n_s}\right)^2 \frac{\gamma_s}{k_x}\right], \qquad \text{(D.26)}$$

where m is a nonnegative integer.

E Characteristic Matrix in Discrete Approach

E.1 Fundamental Equation

Up to now, we have used *MKSA units* in electromagnetism. However, for brevity, we use *CGS-Gaussian units* here.

When the electric current does not flow, *Maxwell's equations* in a uniform optical material with the refractive index n_r and permeability μ are written as

$$\nabla \times \boldsymbol{E} = -\frac{1}{c}\frac{\partial \boldsymbol{H}}{\partial t}, \quad \nabla \times \boldsymbol{H} = \frac{n_r^2}{c}\frac{\partial \boldsymbol{E}}{\partial t}, \tag{E.1}$$

where $\boldsymbol{E}$ is the electric field, c is the speed of light in a vacuum, and $\boldsymbol{H}$ is the magnetic field. We can also express (E.1) with each component as

$$\frac{\partial E_z}{\partial y} - \frac{\partial E_y}{\partial z} = -\frac{1}{c}\frac{\partial H_x}{\partial t}, \quad \frac{\partial H_z}{\partial y} - \frac{\partial H_y}{\partial z} = \frac{n_r^2}{c}\frac{\partial E_x}{\partial t},$$

$$\frac{\partial E_x}{\partial z} - \frac{\partial E_z}{\partial x} = -\frac{1}{c}\frac{\partial H_y}{\partial t}, \quad \frac{\partial H_x}{\partial z} - \frac{\partial H_z}{\partial x} = \frac{n_r^2}{c}\frac{\partial E_y}{\partial t}, \tag{E.2}$$

$$\frac{\partial E_y}{\partial x} - \frac{\partial E_x}{\partial y} = -\frac{1}{c}\frac{\partial H_z}{\partial t}, \quad \frac{\partial H_y}{\partial x} - \frac{\partial H_x}{\partial y} = \frac{n_r^2}{c}\frac{\partial E_z}{\partial t}.$$

We assume that a light propagates toward a positive direction on the z-axis, and dependence of the electric field $\boldsymbol{E}$ and the magnetic field $\boldsymbol{H}$ on time is expressed as $e^{i\omega t}$. In this case, we obtain $\partial/\partial t = i\omega$. Therefore, (E.2) reduces to

$$\frac{\partial E_z}{\partial y} - \frac{\partial E_y}{\partial z} = -\frac{i\omega}{c}H_x, \quad \frac{\partial H_z}{\partial y} - \frac{\partial H_y}{\partial z} = \frac{i\omega n_r^2}{c}E_x,$$

$$\frac{\partial E_x}{\partial z} - \frac{\partial E_z}{\partial x} = -\frac{i\omega}{c}H_y, \quad \frac{\partial H_x}{\partial z} - \frac{\partial H_z}{\partial x} = \frac{i\omega n_r^2}{c}E_y, \tag{E.3}$$

$$\frac{\partial E_y}{\partial x} - \frac{\partial E_x}{\partial y} = -\frac{i\omega}{c}H_z, \quad \frac{\partial H_y}{\partial x} - \frac{\partial H_x}{\partial y} = \frac{i\omega n_r^2}{c}E_z.$$

E.2 TE Mode

We consider the optical waveguide shown in Fig. D.1. Because the lightwave is uniform along the y-axis, we have $\partial/\partial y = 0$, which results in $E_x = E_z = 0$

and $H_y = 0$ for the TE wave from (D.10). Therefore, (E.3) reduces to

$$\frac{\partial E_y}{\partial z} = \frac{i\omega}{c} H_x, \quad \frac{\partial E_y}{\partial x} = -\frac{i\omega}{c} H_z,$$

$$\frac{\partial H_x}{\partial z} - \frac{\partial H_z}{\partial x} = \frac{i\omega n_r^2}{c} E_y. \tag{E.4}$$

From (E.4), we obtain a wave equation as

$$\frac{\partial^2 E_y}{\partial x^2} + \frac{\partial^2 E_y}{\partial z^2} = \frac{i\omega}{c}\left(\frac{\partial H_x}{\partial z} - \frac{\partial H_z}{\partial x}\right) = -\frac{\omega^2 n_r^2}{c^2} E_y. \tag{E.5}$$

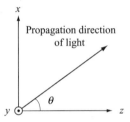

Fig. E.1. Definition of θ

Here, we assume a *separation-of-variables procedure*. Substituting $E_y = X(x)U(z)$ into (E.5) and then dividing both sides by E_y lead to

$$\frac{1}{X}\frac{\partial^2 X}{\partial x^2} + \frac{1}{U}\frac{\partial^2 U}{\partial z^2} = -\frac{\omega^2 n_r^2}{c^2} = -k_0^2 n_r^2, \tag{E.6}$$

where $k_0 = \omega/c$. From (E.6), if we put

$$\frac{1}{X}\frac{\partial^2 X}{\partial x^2} = -k_0^2 n_r^2 \sin^2\theta, \quad \frac{1}{U}\frac{\partial^2 U}{\partial z^2} = -k_0^2 n_r^2 \cos^2\theta, \tag{E.7}$$

where θ is defined in Fig. E.1, we can express E_y as

$$E_y = U(z)\,\exp[\,i(\omega t - k_0 n_r x \sin\theta)\,]. \tag{E.8}$$

Similarly, we have

$$H_x = V(z)\,\exp[\,i(\omega t - k_0 n_r x \sin\theta)\,],$$
$$H_z = W(z)\,\exp[\,i(\omega t - k_0 n_r x \sin\theta)\,]. \tag{E.9}$$

Substituting (E.8) and (E.9) into (E.4) results in

$$\frac{dU}{dz} = \frac{i\omega}{c} V = i k_0 V, \tag{E.10}$$

$$n_r \sin\theta \cdot U = W, \tag{E.11}$$

$$\frac{dV}{dz} + i k_0 n_r \sin\theta \cdot W = i k_0 n_r{}^2 U. \tag{E.12}$$

From (E.11) and (E.12), we obtain

$$\frac{dV}{dz} = i k_0 n_r{}^2 \cos^2\theta \cdot U. \tag{E.13}$$

Differentiating (E.13) with respect to z with the help of (E.10), we have

$$\frac{d^2V}{dz^2} + k_0{}^2 n_r{}^2 \cos^2\theta \cdot V = 0. \tag{E.14}$$

In summary, a relationship between U and V is written as

$$\frac{d^2U}{dz^2} + k_0{}^2 n_r{}^2 \cos^2\theta \cdot U = 0, \quad \frac{d^2V}{dz^2} + k_0{}^2 n_r{}^2 \cos^2\theta \cdot V = 0, \tag{E.15}$$

$$\frac{dU}{dz} = i k_0 V, \quad \frac{dV}{dz} = i k_0 n_r{}^2 \cos^2\theta \cdot U. \tag{E.16}$$

From (E.15) and (E.16), solutions for U and V are expressed as

$$U = A\cos(k_0 n_r z \cos\theta) + B\sin(k_0 n_r z \cos\theta), \tag{E.17}$$

$$V = i n_r \cos\theta [A\sin(k_0 n_r z \cos\theta) - B\cos(k_0 n_r z \cos\theta)]. \tag{E.18}$$

Because (E.15) is a linear differential equation of the second order, we can express U and V as

$$\begin{bmatrix} U(z) \\ V(z) \end{bmatrix} = \begin{bmatrix} F(z) & f(z) \\ G(z) & g(z) \end{bmatrix} \begin{bmatrix} U(0) \\ V(0) \end{bmatrix}, \tag{E.19}$$

where

$$U_1 = f(z), \quad U_2 = F(z), \quad V_1 = g(z), \quad V_2 = G(z). \tag{E.20}$$

Also, from (E.16), we have the following relation

$$\frac{dU_1}{dz} \equiv U_1' = i k_0 V_1, \quad \frac{dU_2}{dz} \equiv U_2' = i k_0 V_2, \tag{E.21}$$

$$\frac{dV_1}{dz} \equiv V_1' = i k_0 n_r{}^2 \cos^2\theta \cdot U_1, \quad \frac{dV_2}{dz} \equiv V_2' = i k_0 n_r{}^2 \cos^2\theta \cdot U_2, \tag{E.22}$$

from which we obtain

$$U_1 V_2' - V_1' U_2 = 0, \quad V_1 U_2' - U_1' V_2 = 0. \tag{E.23}$$

As a result, we have

$$\frac{\mathrm{d}}{\mathrm{d}z}(U_1 V_2 - V_1 U_2) = 0. \tag{E.24}$$

Therefore, the following determinant is satisfied:

$$\begin{vmatrix} U_1 & U_2 \\ V_1 & V_2 \end{vmatrix} = \text{constant}. \tag{E.25}$$

As the matrix elements satisfying (E.25), we consider

$$f(0) = G(0) = 0, \quad F(0) = g(0) = 1,$$
$$F(z)g(z) - f(z)G(z) = 1. \tag{E.26}$$

From (E.19) and (E.26), we define the *transfer matrix* M as

$$\begin{bmatrix} U(0) \\ V(0) \end{bmatrix} = M \begin{bmatrix} U(z) \\ V(z) \end{bmatrix}, \quad M = \begin{bmatrix} g(z) & -f(z) \\ -G(z) & F(z) \end{bmatrix}. \tag{E.27}$$

Also, from (E.17), (E.18), and (E.26), we put

$$f(z) = \frac{\mathrm{i}}{n_r \cos \theta} \sin(k_0 n_r z \cos \theta),$$

$$F(z) = \cos(k_0 n_r z \cos \theta),$$

$$g(z) = \cos(k_0 n_r z \cos \theta), \tag{E.28}$$

$$G(z) = \mathrm{i}\, n_r \cos \theta \cdot \sin(k_0 n_r z \cos \theta).$$

If we place

$$\beta_i = k_0 n_r z \cos \theta, \quad p_i = n_r \cos \theta, \tag{E.29}$$

and substitute (E.28) and (E.29) into (E.27), we can express the transfer matrix M as

$$M = \begin{bmatrix} \cos \beta_i & -\dfrac{\mathrm{i}}{p_i} \sin \beta_i \\ -\mathrm{i}\, p_i \sin \beta_i & \cos \beta_i \end{bmatrix} = \begin{bmatrix} m_{11} & m_{12} \\ m_{21} & m_{22} \end{bmatrix}. \tag{E.30}$$

As shown in Fig. E.2, we consider a light propagating along the arrow in sequentially placed layers with different refractive indexes. The tangent of the electric field E_y and that of the magnetic field H_x are continuous at each interface. Therefore, if we write the tangent components of the electric fields for the incident wave, the reflected wave, and the transmitted wave as A, R, and T, respectively, the boundary condition is expressed as

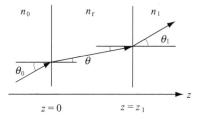

Fig. E.2. Propagation of light in a region with different refractive indexes

$$A + R = U(0), \quad T = U(z_1), \tag{E.31}$$
$$p_0(A - R) = V(0), \quad p_1 T = V(z_1), \tag{E.32}$$

where

$$p_0 = n_0 \cos\theta, \quad p_1 = n_1 \cos\theta. \tag{E.33}$$

Here, n_0 and n_1 are the refractive indexes of the first and last media, respectively. From (E.27) and (E.30)–(E.33), we obtain

$$A + R = (m_{11} + m_{12}p_1)T, \tag{E.34}$$
$$p_0(A - R) = (m_{21} + m_{22}p_1)T. \tag{E.35}$$

From (E.34) and (E.35), the amplitude reflectivity r and the amplitude transmissivity t are given by

$$r = \frac{R}{A} = \frac{(m_{11} + m_{12}p_1)p_0 - (m_{21} + m_{22}p_1)}{(m_{11} + m_{12}p_1)p_0 + (m_{21} + m_{22}p_1)}, \tag{E.36}$$
$$t = \frac{T}{A} = \frac{2p_0}{(m_{11} + m_{12}p_1)p_0 + (m_{21} + m_{22}p_1)}. \tag{E.37}$$

Using the amplitude reflectivity r and the amplitude transmissivity t, the power reflectivity R and the power transmissivity T are expressed as

$$R = r^*r, \quad T = \frac{p_1}{p_0} t^*t. \tag{E.38}$$

E.3 TM Mode

For the TM wave, we have $E_y = 0$ and $H_x = H_z = 0$ from (D.22). Other components of the TM wave are written as

$$\begin{aligned}
H_y &= U(z) \exp[\mathrm{i}\,(\omega t - k_0 n_r x \sin\theta)], \\
E_x &= -V(z) \exp[\mathrm{i}\,(\omega t - k_0 n_r x \sin\theta)], \\
E_z &= -W(z) \exp[\mathrm{i}\,(\omega t - k_0 n_r x \sin\theta)].
\end{aligned} \tag{E.39}$$

Also, from (E.3), we have

$$\frac{\partial H_y}{\partial z} = -\frac{i\omega n_r^2}{c} E_x = -i k_0 n_r^2 E_x,$$

$$\frac{\partial H_y}{\partial x} = \frac{i\omega n_r^2}{c} E_z = i k_0 n_r^2 E_z, \tag{E.40}$$

$$\frac{\partial E_x}{\partial z} - \frac{\partial E_z}{\partial x} = -\frac{i\omega}{c} H_y = -i k_0 H_y.$$

Therefore, we obtain a wave equation

$$\frac{\partial^2 H_y}{\partial x^2} + \frac{\partial^2 H_y}{\partial z^2} = -i k_0 n_r^2 \left(\frac{\partial E_x}{\partial z} - \frac{\partial E_z}{\partial x} \right) = -k_0 n_r^2 H_y. \tag{E.41}$$

Substituting (E.39) into (E.41) gives

$$\frac{d^2 U}{dz^2} + (k_0 n_r \cos \theta)^2 U = 0. \tag{E.42}$$

Inserting (E.39) into (E.40) results in

$$\frac{dU}{dz} = i k_0 n_r^2 V, \tag{E.43}$$

$$\sin \theta \cdot U = n_r W, \tag{E.44}$$

$$\frac{dV}{dz} + i k_0 n_r \sin \theta \cdot W = i k_0 U. \tag{E.45}$$

From (E.44) and (E.45), we obtain

$$\frac{dV}{dz} = i k_0 \cos^2 \theta \cdot U. \tag{E.46}$$

Differentiating (E.46) with respect to z with the help of (E.43) results in

$$\frac{d^2 V}{dz^2} + (k_0 n_r \cos \theta)^2 V = 0. \tag{E.47}$$

In summary, a relationship between U and V for the TM mode is written as

$$\frac{d^2 U}{dz^2} + k_0^2 n_r^2 \cos^2 \theta \cdot U = 0, \quad \frac{d^2 V}{dz^2} + k_0^2 n_r^2 \cos^2 \theta \cdot V = 0, \tag{E.48}$$

$$\frac{dU}{dz} = i k_0 n_r^2 V, \quad \frac{dV}{dz} = i k_0 \cos^2 \theta \cdot U. \tag{E.49}$$

For the TM mode, the transfer matrix $\boldsymbol{M}$, the amplitude reflectivity r, the amplitude transmissivity t, the power reflectivity R, and the power transmissivity T are obtained by replacing p_i for the TE mode with

$$q_i = \frac{\cos \theta}{n_r}. \tag{E.50}$$

F Free Carrier Absorption and Plasma Effect

An equation of motion for the electron in a crystal without a magnetic field is given by

$$m^* \frac{\mathrm{d}^2 x}{\mathrm{d}t^2} + \frac{1}{\tau} m^* \frac{\mathrm{d}x}{\mathrm{d}t} = -eE, \tag{F.1}$$

where m^* is the effective mass of the electron, τ is the relaxation time such as the mean free time of collision, e is the elementary charge, and E is the electric field. If we assume x, $E \propto \mathrm{e}^{\mathrm{i}\omega t}$ where ω is the angular frequency of a light, a position x of the electron is obtained as

$$x = \frac{eE}{m^* \left(\omega^2 - \mathrm{i}\omega/\tau\right)}. \tag{F.2}$$

The polarization P of a material is written as

$$P = P_0 + P_1; \quad P_1 = -nex, \tag{F.3}$$

where P_0 is the polarization caused by ionization of atoms constituting the crystal, and P_1 is the polarization induced by the motion of the electrons. Here, we express the electric flux density D as

$$D = \varepsilon_0 E + P = (\varepsilon_0 E + P_0) + P_1 = \varepsilon \varepsilon_0 E + P_1 = \varepsilon' \varepsilon_0 E, \tag{F.4}$$

where ε_0 is permittivity in a vacuum and ε is the dielectric constant of the crystal based on ionization of the atoms. The dielectric constant ε', which is modified by the motion of the electrons, is written as

$$\varepsilon' = \varepsilon_\mathrm{r} - \mathrm{i}\varepsilon_\mathrm{i} = \varepsilon - \frac{nex}{\varepsilon_0 E}, \tag{F.5}$$

where ε_r and ε_i are real and imaginary parts of ε', respectively. Substituting (F.2) into (F.5) leads to

$$\varepsilon_\mathrm{r} - \varepsilon \simeq -\frac{ne^2}{m^* \omega^2 \varepsilon_0}, \tag{F.6}$$

$$\varepsilon_\mathrm{i} \simeq \frac{ne^2}{m^* \omega^3 \varepsilon_0 \tau}, \tag{F.7}$$

where $\omega \gg 1/\tau$ was used.

Using the complex refractive index $n_r - i\kappa$, we can express ε_r and ε_i as

$$\varepsilon_r = n_r{}^2 - \kappa^2, \tag{F.8}$$

$$\varepsilon_i = 2n_r\kappa. \tag{F.9}$$

Therefore, the optical power absorption coefficient α due to free carrier absorption is obtained as

$$\alpha = \frac{2\omega}{c}\kappa = \frac{\omega}{c}\frac{\varepsilon_i}{n_r} = \frac{ne^2}{m^*\omega^2\varepsilon_0 n_r c\tau}, \tag{F.10}$$

where c is the speed of light in a vacuum.

When the carrier concentration increases by n, we assume that ε_r and n_r change to $\varepsilon_r + \Delta\varepsilon_r$ and $n_r + \Delta n_r$, respectively. In this case, we have

$$2n_r\Delta n_r = \Delta\varepsilon_r. \tag{F.11}$$

Hence, we obtain

$$\Delta n_r = \frac{\Delta\varepsilon_r}{2n_r} = \frac{\varepsilon_r - \varepsilon}{2n_r} = -\frac{e^2}{2m^*\omega^2\varepsilon_0 n_r}n, \tag{F.12}$$

which is referred to as the free carrier plasma effect.

G Relative Intensity Noise (RIN)

G.1 Rate Equations with Fluctuations

When semiconductor lasers show two-mode operations, the rate equations on the photon densities S_1 and S_2 and the carrier concentration n are expressed as

$$\frac{dS_1}{dt} = (\alpha_1 - \beta_1 S_1 - \theta_{12} S_2)\, S_1 + \beta_{\mathrm{sp1}}\frac{n}{\tau_n} + F_1(t), \tag{G.1}$$

$$\frac{dS_2}{dt} = (\alpha_2 - \beta_2 S_2 - \theta_{21} S_1)\, S_2 + \beta_{\mathrm{sp2}}\frac{n}{\tau_n} + F_2(t), \tag{G.2}$$

$$\frac{dn}{dt} = \frac{I}{eV_{\mathrm{A}}} - [G_1(n) - \beta_1 S_1 - \theta_{12} S_2]\, S_1$$
$$- [G_2(n) - \beta_2 S_2 - \theta_{21} S_1]\, S_2 - \frac{n}{\tau_n} + F_n(t). \tag{G.3}$$

Here, $\alpha_i \equiv G_i(n) - 1/\tau_{\mathrm{phi}}$ is the net amplification rate, $G_i(n)$ is the amplification rate, τ_{phi} is the photon lifetime, β_i is the self-saturation coefficient, θ_{ij} is the cross-saturation coefficient, $\beta_{\mathrm{sp}i}$ is the spontaneous emission coupling factor $(i, j = 1, 2)$, τ_n is the carrier lifetime, I is the injection current, e is the elementary charge, and V_{A} is a volume of the active layer.

Fluctuations are expressed by the Langevin noise sources F_1, F_2, and F_n. In Markoffian approximation, time averages of F_1, F_2, and F_n satisfy

$$\langle F_k(t) \rangle = 0, \tag{G.4}$$

$$\langle F_k(t) F_l(t') \rangle = 2D_{kl}\delta(t - t'), \tag{G.5}$$

where k, $l = 1$, 2, and n. The diffusion coefficients D_{kl} are given by

$$D_{11} = \beta_{\mathrm{sp1}}\frac{n}{\tau_n} S_{10}, \quad D_{22} = \beta_{\mathrm{sp2}}\frac{n}{\tau_n} S_{20}, \quad D_{12} = D_{21} = 0,$$

$$D_{nn} = \beta_{\mathrm{sp1}}\frac{n}{\tau_n} S_{10} + \beta_{\mathrm{sp2}}\frac{n}{\tau_n} S_{20} + \frac{n}{\tau_n}, \tag{G.6}$$

$$D_{1n} = -\beta_{\mathrm{sp1}}\frac{n}{\tau_n} S_{10}, \quad D_{2n} = -\beta_{\mathrm{sp2}}\frac{n}{\tau_n} S_{20},$$

where S_{10} and S_{20} are the average photon densities of mode 1 and mode 2, respectively.

The amplification rate $G_i(n)$ is given by

$$G_i(n) = \left[\frac{\partial G_i}{\partial n}\right]_{n=n_{\text{thi}}} (n - n_{0i}), \tag{G.7}$$

where n_{thi} is the threshold carrier concentration and n_{0i} is the transparent carrier concentration.

G.2 RIN without Carrier Fluctuations

We express the photon densities S_1 and S_2 as

$$S_1 = S_{10} + \delta S_1(t), \quad S_2 = S_{20} + \delta S_2(t), \tag{G.8}$$

where S_{10} and S_{20} are the average photon densities in a steady state, and δS_1 and δS_2 are the fluctuations of mode 1 and mode 2, respectively. Substituting Fourier transforms

$$\delta S_i(t) = \frac{1}{2\pi} \int_{-\infty}^{\infty} \delta \tilde{S}_i(\omega) \exp(\mathrm{i}\omega t)\, \mathrm{d}\omega, \tag{G.9}$$

$$\delta \tilde{S}_i(\omega) = \int_{-\infty}^{\infty} \delta S_i(t) \exp(-\mathrm{i}\omega t)\, \mathrm{d}t, \tag{G.10}$$

$$F_i(t) = \frac{1}{2\pi} \int_{-\infty}^{\infty} \tilde{F}_i(\omega) \exp(\mathrm{i}\omega t)\, \mathrm{d}\omega, \tag{G.11}$$

$$\tilde{F}_i(\omega) = \int_{-\infty}^{\infty} F_i(t) \exp(-\mathrm{i}\omega t)\, \mathrm{d}t, \tag{G.12}$$

into (G.1)–(G.3), and then neglecting $F_n(t)$, we have

$$\mathrm{i}\omega \delta \tilde{S}_1(\omega) = -A_1 \delta \tilde{S}_1(\omega) - B_1 \delta \tilde{S}_2(\omega) + \tilde{F}_1(\omega), \tag{G.13}$$

$$\mathrm{i}\omega \delta \tilde{S}_2(\omega) = -A_2 \delta \tilde{S}_2(\omega) - B_2 \delta \tilde{S}_1(\omega) + \tilde{F}_2(\omega), \tag{G.14}$$

where

$$A_1 = \beta_1 S_{10}, \quad A_2 = \beta_2 S_{20}, \tag{G.15}$$

$$B_1 = \theta_{12} S_{10}, \quad B_2 = \theta_{21} S_{20}. \tag{G.16}$$

From (G.13) and (G.14), the self-correlation functions $\langle \delta \tilde{S}_1(\omega) \delta \tilde{S}_1^{*}(\omega') \rangle$ and $\langle \delta \tilde{S}_2(\omega) \delta \tilde{S}_2^{*}(\omega') \rangle$ are given by

$$\langle \delta \tilde{S}_1(\omega) \delta \tilde{S}_1^{*}(\omega') \rangle = \tilde{S}_{\delta S_1}(\omega) \cdot 2\pi \delta(\omega - \omega'), \tag{G.17}$$

$$\tilde{S}_{\delta S_1}(\omega) = \frac{(\omega^2 + A_2^2)W_1 - 2A_2 B_1 W_{12} + B_1^2 W_2}{(A_1 A_2 - B_1 B_2 - \omega^2)^2 + \omega^2(A_1 + A_2)^2}, \tag{G.18}$$

$$\langle \delta \tilde{S}_2(\omega) \delta \tilde{S}_2^{*}(\omega') \rangle = \tilde{S}_{\delta S_2}(\omega) \cdot 2\pi \delta(\omega - \omega'), \tag{G.19}$$

$$\tilde{S}_{\delta S_2}(\omega) = \frac{(\omega^2 + A_1^2)W_2 - 2A_1 B_2 W_{21} + B_2^2 W_1}{(A_1 A_2 - B_1 B_2 - \omega^2)^2 + \omega^2(A_1 + A_2)^2}, \tag{G.20}$$

where

$$W_1 = \langle \tilde{F}_1(\omega)\tilde{F}_1^*(\omega)\rangle, \tag{G.21}$$

$$W_2 = \langle \tilde{F}_2(\omega)\tilde{F}_2^*(\omega)\rangle, \tag{G.22}$$

$$W_{12} = \langle \tilde{F}_1(\omega)\tilde{F}_2^*(\omega)\rangle. \tag{G.23}$$

In the shot noise Langevin model, the correlation strengths $\langle \tilde{F}_k(\omega)\tilde{F}_k^*(\omega)\rangle$ and $\langle \tilde{F}_k(\omega)\tilde{F}_l^*(\omega)\rangle$ are given by

$$\langle \tilde{F}_k(\omega)\tilde{F}_k^*(\omega)\rangle, = \sum R_k^+ + \sum R_k^- \tag{G.24}$$

$$\langle \tilde{F}_k(\omega)\tilde{F}_l^*(\omega)\rangle = -\left(\sum R_{kl} + \sum R_{lk}\right), \tag{G.25}$$

where R_k^+ and R_k^- ($k = 1, 2$) are the rate for the photon to enter the photon reservoirs and to leave them, respectively; R_k^+ and R_k^- ($k = n$) are the rate for the electron to enter the electron reservoirs and to leave them, respectively; and R_{kl} and R_{lk} are the exchange rates.

As a result, the correlation strength $\langle \tilde{F}_1(\omega)\tilde{F}_1^*(\omega)\rangle$ is obtained as

$$\begin{aligned} W_1 &= \langle \tilde{F}_1(\omega)\tilde{F}_1^*(\omega)\rangle \\ &= \frac{1}{V_A}\left[\left(\alpha_1 + \frac{1}{\tau_{ph1}} + \beta_1 S_{10} + \theta_{12}S_{20}\right)S_{10} + \beta_{sp1}\frac{n}{\tau_n}\right] \\ &\simeq 2\left(\alpha_1 + \frac{1}{\tau_{ph1}}\right)\frac{S_{10}}{V_A}. \end{aligned} \tag{G.26}$$

Similarly, we have

$$W_2 = \langle \tilde{F}_2(\omega)\tilde{F}_2^*(\omega)\rangle \simeq 2\left(\alpha_2 + \frac{1}{\tau_{ph2}}\right)\frac{S_{20}}{V_A}, \tag{G.27}$$

$$W_{12} = \langle \tilde{F}_1(\omega)\tilde{F}_2^*(\omega)\rangle \simeq (\theta_{12} + \theta_{21})\frac{S_{10}S_{20}}{V_A} = W_{21}. \tag{G.28}$$

Using (G.18) and (G.20), we define the *relative intensity noises (RINs)* per unit bandwidth as

$$\mathrm{RIN}_1 = \frac{2\tilde{S}_{\delta S_1}(\omega)}{S_{10}^2} = \frac{2}{S_{10}^2}\frac{(\omega^2 + A_2{}^2)W_1 - 2A_2B_1W_{12} + B_1{}^2W_2}{(A_1A_2 - B_1B_2 - \omega^2)^2 + \omega^2(A_1 + A_2)^2}, \tag{G.29}$$

$$\mathrm{RIN}_2 = \frac{2\tilde{S}_{\delta S_2}(\omega)}{S_{20}^2} = \frac{2}{S_{20}^2}\frac{(\omega^2 + A_1{}^2)W_2 - 2A_1B_2W_{21} + B_2{}^2W_1}{(A_1A_2 - B_1B_2 - \omega^2)^2 + \omega^2(A_1 + A_2)^2}. \tag{G.30}$$

G.3 RIN with Carrier Fluctuations

We express the carrier concentration n as

$$n = n_{c0} + \delta n(t), \tag{G.31}$$

where n_{c0} is the carrier concentration in a steady state and δn is the fluctuation in the carrier concentration. Their Fourier transforms are written as

$$\delta n(t) = \frac{1}{2\pi} \int_{-\infty}^{\infty} \delta \tilde{n}(\omega) \exp(\mathrm{i}\,\omega t)\,\mathrm{d}\omega, \tag{G.32}$$

$$\delta \tilde{n}(\omega) = \int_{-\infty}^{\infty} \delta n(t) \exp(-\mathrm{i}\,\omega t)\,\mathrm{d}t, \tag{G.33}$$

$$F_n(t) = \frac{1}{2\pi} \int_{-\infty}^{\infty} \tilde{F}_n(\omega) \exp(\mathrm{i}\,\omega t)\,\mathrm{d}\omega, \tag{G.34}$$

$$\tilde{F}_n(\omega) = \int_{-\infty}^{\infty} F_n(t) \exp(-\mathrm{i}\,\omega t)\,\mathrm{d}t. \tag{G.35}$$

Substituting (G.9)–(G.12) and (G.32)–(G.35) into (G.1)–(G.3), we have

$$\mathrm{i}\,\omega\delta\tilde{S}_1(\omega) = -A_1\delta\tilde{S}_1(\omega) - B_1\delta\tilde{S}_2(\omega) + C_1\delta\tilde{n}(\omega) + \tilde{F}_1(\omega), \tag{G.36}$$

$$\mathrm{i}\,\omega\delta\tilde{S}_2(\omega) = -A_2\delta\tilde{S}_2(\omega) - B_2\delta\tilde{S}_1(\omega) + C_2\delta\tilde{n}(\omega) + \tilde{F}_2(\omega), \tag{G.37}$$

$$\mathrm{i}\,\omega\delta\tilde{n}(\omega) = -G_1(n_{c0})\delta\tilde{S}_1(\omega) - G_2(n_{c0})\delta\tilde{S}_2(\omega)$$
$$+ C_3\delta\tilde{n}(\omega) + \tilde{F}_n(\omega), \tag{G.38}$$

where

$$C_1 = \left[\frac{\partial G_1}{\partial n}\right]_{n=n_{\mathrm{th}}} S_{10} + \frac{\beta_{\mathrm{sp1}}}{\tau_n}, \tag{G.39}$$

$$C_2 = \left[\frac{\partial G_2}{\partial n}\right]_{n=n_{\mathrm{th}}} S_{20} + \frac{\beta_{\mathrm{sp2}}}{\tau_n}, \tag{G.40}$$

$$C_3 = C_1 + C_2 + \frac{1}{\tau_n}. \tag{G.41}$$

From (G.36)–(G.38), the self-correlation functions $\langle \delta\tilde{S}_1(\omega)\delta\tilde{S}_1{}^*(\omega')\rangle$ and $\langle \delta\tilde{S}_2(\omega)\delta\tilde{S}_2{}^*(\omega')\rangle$ are given by

$$\langle \delta \tilde{S}_1(\omega) \delta \tilde{S}_1^{\,*}(\omega') \rangle = \tilde{S}_{\delta S_1}(\omega) \cdot 2\pi \delta(\omega - \omega'), \tag{G.42}$$

$$\tilde{S}_{\delta S_1}(\omega) = \frac{1}{|X_1 X_2 - Y_1 Y_2|^2}$$
$$\times \{ |X_2|^2 W_1 - (X_2^* Y_1 + X_2 Y_1^*) W_{12}$$
$$+ |Y_1|^2 W_2 + |Z_1 X_2 - Z_2 Y_1|^2 W_n$$
$$+ [X_2^*(Z_1 X_2 - Z_2 Y_1) + X_2(Z_1 X_2 - Z_2 Y_1)^*] W_{1n}$$
$$- [Y_1^*(Z_1 X_2 - Z_2 Y_1) + Y_1(Z_1 X_2 - Z_2 Y_1)^*] W_{2n} \}, \tag{G.43}$$

$$\langle \delta \tilde{S}_2(\omega) \delta \tilde{S}_2^{\,*}(\omega') \rangle = \tilde{S}_{\delta S_2}(\omega) \cdot 2\pi \delta(\omega - \omega'), \tag{G.44}$$

$$\tilde{S}_{\delta S_2}(\omega) = \frac{1}{|X_1 X_2 - Y_1 Y_2|^2}$$
$$\times \{ |Y_2|^2 W_1 - (X_1^* Y_2 + X_1 Y_2^*) W_{12}$$
$$+ |X_1|^2 W_2 + |Z_2 X_1 - Z_1 Y_2|^2 W_n$$
$$- [Y_2^*(Z_2 X_1 - Z_1 Y_2) + Y_2(Z_2 X_1 - Z_1 Y_2)^*] W_{1n}$$
$$+ [X_1^*(Z_2 X_1 - Z_1 Y_2) + X_1(Z_2 X_1 - Z_1 Y_2)^*] W_{2n} \},$$
$$\tag{G.45}$$

where

$$X_1 = \mathrm{i}\,\omega + A_1 + \frac{C_1 G_1(n_{c0})}{\mathrm{i}\,\omega + C_3}, \tag{G.46}$$

$$X_2 = \mathrm{i}\,\omega + A_2 + \frac{C_2 G_2(n_{c0})}{\mathrm{i}\,\omega + C_3}, \tag{G.47}$$

$$Y_1 = B_1 + \frac{C_1 G_2(n_{c0})}{\mathrm{i}\,\omega + C_3}, \tag{G.48}$$

$$Y_2 = B_2 + \frac{C_2 G_1(n_{c0})}{\mathrm{i}\,\omega + C_3}, \tag{G.49}$$

$$Z_1 = \frac{C_1}{\mathrm{i}\,\omega + C_3}, \tag{G.50}$$

$$Z_2 = \frac{C_2}{\mathrm{i}\,\omega + C_3}, \tag{G.51}$$

$$W_n = \langle \tilde{F}_n(\omega) \tilde{F}_n^{\,*}(\omega) \rangle = \frac{2I_0}{e V_A^{\,2}}, \tag{G.52}$$

$$W_{1n} = \langle \tilde{F}_1(\omega) \tilde{F}_n^{\,*}(\omega) \rangle = -W_1, \tag{G.53}$$

$$W_{2n} = \langle \tilde{F}_2(\omega) \tilde{F}_n^{\,*}(\omega) \rangle = -W_2. \tag{G.54}$$

Using (G.43) and (G.45), we obtain the RINs per unit bandwidth as

$$\text{RIN}_1 = \frac{2\tilde{S}_{\delta S_1}(\omega)}{S_{10}{}^2}$$

$$= \frac{2}{S_{10}{}^2} \frac{1}{|X_1 X_2 - Y_1 Y_2|^2}$$
$$\times \{|X_2|^2 W_1 - (X_2{}^* Y_1 + X_2 Y_1{}^*) W_{12}$$
$$+ |Y_1|^2 W_2 + |Z_1 X_2 - Z_2 Y_1|^2 W_n$$
$$+ [X_2{}^*(Z_1 X_2 - Z_2 Y_1) + X_2 (Z_1 X_2 - Z_2 Y_1)^*] W_{1n}$$
$$- [Y_1{}^*(Z_1 X_2 - Z_2 Y_1) + Y_1 (Z_1 X_2 - Z_2 Y_1)^*] W_{2n}\}, \qquad \text{(G.55)}$$

$$\text{RIN}_2 = \frac{2\tilde{S}_{\delta S_2}(\omega)}{S_{20}{}^2}$$

$$= \frac{2}{S_{20}{}^2} \frac{1}{|X_1 X_2 - Y_1 Y_2|^2}$$
$$\times \{|Y_2|^2 W_1 - (X_1{}^* Y_2 + X_1 Y_2{}^*) W_{12}$$
$$+ |X_1|^2 W_2 + |Z_2 X_1 - Z_1 Y_2|^2 W_n$$
$$- [Y_2{}^*(Z_2 X_1 - Z_1 Y_2) + Y_2 (Z_2 X_1 - Z_1 Y_2)^*] W_{1n}$$
$$+ [X_1{}^*(Z_2 X_1 - Z_1 Y_2) + X_1 (Z_2 X_1 - Z_1 Y_2)^*] W_{2n}\}. \qquad \text{(G.56)}$$

References

1. E. O. Kane: "Band structure of indium antimonide," J. Phys. Chem. Solids **1**, 249 (1957)
2. C. Kittel: *Quantum Theory of Solids*, 2nd edn (John Wiley & Sons, New York 1987)
3. G. Bastard: *Wave Mechanics Applied to Semiconductor Heterostructures* (Halsted Press, New York 1988)
4. S. Datta: *Quantum Phenomena* (Addison-Wesley, Reading 1989)
5. L. I. Schiff: *Quantum Mechanics*, 3rd edn (McGraw-Hill, New York 1968)
6. A. Messiah: *Quantum Mechanics*, vol.1, vol.2 (North-Holland, Amsterdam 1961)
7. A. Einstein: "Quantentheorie der Strahlung," Phys. Z. **18**, 121 (1917)
8. J. I. Pankove: *Optical Processes in Semiconductors* (Dover, New York 1975)
9. T. Tamir, ed.: *Integrated Optics*, 2nd edn (Springer, Berlin 1979)
10. T. Tamir, ed.: *Guided-Wave Optoelectronics*, 2nd edn (Springer, Berlin 1990)
11. R. G. Hunsperger: *Integrated Optics: Theory and Technology*, 3rd edn (Springer, Berlin 1991)
12. D. Marcuse: *Theory of Dielectric Waveguides*, 2nd edn (Academic Press, San Diego 1991)
13. K. J. Ebeling: *Integrated Optoelectronics* (Springer, Berlin 1992)
14. E. A. J. Marcatili: "Dielectric rectangular waveguide and directional coupler for integrated optics," Bell Syst. Tech. J. **48**, 2071 (1969)
15. T. Numai: "1.5 µm phase-shift-controlled distributed feedback wavelength tunable optical filter," IEEE J. Quantum Electron. **QE-28**, 1513 (1992)
16. H. Kogelnik and C. V. Shank: "Coupled-wave theory of distributed feedback lasers," J. Appl. Phys. **43**, 2327 (1972)
17. M. Yamada and K. Sakuda: "Analysis of almost-periodic distributed feedback slab waveguides via a fundamental matrix approach," Appl. Opt. **26**, 3474 (1987)
18. H. A. Haus and C. V. Shank: "Antisymmetric taper of distributed feedback lasers," IEEE J. Quantum Electron. **QE-12**, 532 (1976)
19. H. A. Haus: *Waves and Fields in Optoelectronics* (Prentice-Hall, Englewood Cliffs 1984)
20. T. Numai: A study on semiconductor wavelength tunable optical filters and lasers. Ph.D. Thesis, Keio University, Yokohama (1992)
21. T. Numai: "1.5-µm wavelength tunable phase-shift-controlled distributed feedback laser," IEEE/OSA J. Lightwave Technol. **10** 199 (1992)
22. K. Utaka, S. Akiba, K. Sakai, et al.: "$\lambda/4$-shifted InGaAsP/InP DFB lasers by simultaneous holographic exposure of positive and negative photoresists," Electron. Lett. **20**, 1008 (1984)

23. K. Utaka, S. Akiba, K. Sakai, et al.: "$\lambda/4$-shifted InGaAsP/InP DFB lasers," IEEE J. Quantum Electron. **QE-22**, 1042 (1986)

24. M. Okai, S. Tsuji, M. Hirao, et al.: "New high resolution positive and negative photoresist method for $\lambda/4$-shifted DFB lasers," Electron. Lett. **23**, 370 (1987)

25. M. Shirasaki, H. Soda, S. Yamakoshi, et al.: "$\lambda/4$-shifted DFB-LD corrugation formed by a novel spatial phase modulating mask," European Conf. Opt. Commun./Integrated Opt. and Opt. Commun. 25 (1985)

26. T. Numai, M. Yamaguchi, I. Mito, et al.: "A new grating fabrication method for phase-shifted DFB LDs," Jpn. J. Appl. Phys., Part 2 **26**, L1910 (1987)

27. S. Tsuji, A. Ohishi, M. Okai, et al.: "Quarter lambda shift DFB lasers by phase image projection method," 10th Int. Laser Conf. 58 (1986)

28. Y. Ono, S. Takano, I. Mito, et al.: "Phase-shifted diffraction-grating fabrication using holographic wavefront reconstruction," Electron. Lett. **23**, 57 (1987)

29. M. Okai, S. Tsuji, N. Chinone, et al.: "Novel method to fabricate corrugation for a $\lambda/4$-shifted distributed feedback laser using a grating photomask," Appl. Phys. Lett. **55**, 415 (1989)

30. H. Sugimoto, Y. Abe, T. Matsui, et al.: "Novel fabrication method of quarter-wave-shifted gratings using ECR-CVD SiN_x films," Electron. Lett. **23**, 1260 (1987)

31. K. Sekartedjo, N. Eda, K. Furuya, et al.: "1.5-μm phase-shifted DFB lasers for single-mode operation," Electron. Lett. **20**, 80 (1984)

32. T. Nishida, M. Nakao, T. Tamamura, et al.: "Synchrotron radiation lithography for DFB laser gratings," Jpn. J. Appl. Phys. **28**, 2333 (1989)

33. Z. I. Alferov, V. M. Andreev, E. L. Portnoy, et al.: "AlAs-GaAs heterojunction injection lasers with a low room-temperature threshold," Sov. Phys. Semicond. **3**, 1107 (1970)

34. I. Hayashi, M. B. Panish, P. W. Foy, et al.: "Junction lasers which operate continuously at room temperature," Appl. Phys. Lett. **17**, 109 (1970)

35. W. E. Lamb, Jr.: "Theory of an optical maser," Phys. Rev. **134**, A1429 (1964)

36. W. H. Louisell: *Quantum Statistical Properties of Radiation* (John Wiley & Sons, New York 1973)

37. M. Sargent III, M. O. Scully, and W. E. Lamb, Jr.: *Laser Physics* (Addison-Wesley, Reading 1974)

38. H. Haken: *Laser Theory* (Springer, Berlin 1984)

39. K. Shimoda: *Introduction to Laser Physics*, 2nd edn (Springer, Berlin 1986)

40. A. E. Siegman: *Lasers* (University Science Books, Mill Valley 1986)

41. A. Yariv: *Quantum Electronics*, 3rd edn (John Wiley & Sons, New York 1989)

42. R. Lang and K. Kobayashi: "External optical feedback effects on semiconductor injection laser properties," IEEE J. Quantum Electron. **QE-16**, 347 (1980)

43. Y. Nakano, Y. Luo, and K. Tada: "Facet reflection independent, single longitudinal mode oscillation in a GaAlAs/GaAs distributed feedback laser equipped with a gain-coupling mechanism," Appl. Phys. Lett. **55**, 1606 (1989)

44. K. Iga, F. Koyama, and S. Kinoshita: "Surface emitting semiconductor lasers," IEEE J. Quantum Electron. **QE-24**, 1845 (1988)

45. Z. L. Liau and J. N. Walpole: " Large monolithic two-dimensional arrays of GaInAsP/InP surface-emitting lasers," Appl. Phys. Lett. **50**, 528 (1987)

46. N. W. Carlson, G. A. Evans, D. P. Bour, et al.: "Demonstration of a grating-surface-emitting diode laser with low-threshold current density ," Appl. Phys. Lett. **56**, 16 (1990)

47. Y. Arakawa and A. Yariv: "Quantum well lasers—gain, spectra, dynamics," IEEE J. Quantum Electron. **QE-22**, 1887 (1986)
48. P. S. Zory, Jr., ed.: *Quantum Well Lasers* (Academic Press, San Diego 1993)
49. M. Okai, M. Suzuki, and T. Taniwatari: "Strained multiquantum-well corrugation-pitch-modulated distributed feedback laser with ultranarrow (3.6 kHz) spectral linewidth," Electron. Lett. **29**, 1696 (1993)
50. J. M. Luttinger and W. Kohn: "Motion of electrons and holes in perturbed periodic fields," Phys. Rev. **97**, 869 (1955)
51. G. E. Pikus and G. L. Bir: "Effect of deformation on the energy spectrum and the electrical properties of imperfect germanium and silicon," Sov. Phys. Solid State **1**, 136 (1959)
52. P. Y. Yu and M. Cardona: *Fundamentals of Semiconductors, Physics and Materials Properties*, 2nd edn (Springer, Berlin 1999)
53. S. Adachi: *Physical Properties of III–V Semiconductor Compounds* (John Wiley & Sons, New York 1992)
54. C. Kittel: *Introduction to Solid State Physics*, 7th edn (John Wiley & Sons, New York 1996)
55. J. F. Nye: *Physical Properties of Crystals* (Oxford University Press, New York 1985)
56. H. Yokoyama and K. Ujihara, eds.: *Spontaneous Emission and Laser Oscillation in Microcavities* (CRC Press, New York 1995)
57. J. Dalibard, J. Dupont-Roc, and C. Cohen-Tannoudji: "Vacuum fluctuations and radiation reaction: identification of their respective contributions," J. Phys. **43**, 1617 (1982)
58. F. Stern and J. M. Woodall: "Photon recycling in semiconductor lasers," J. Appl. Phys. **45**, 3904 (1974)
59. T. Numai, H. Kosaka, I. Ogura, et al.: "Indistinct threshold laser operation in a pnpn vertical to surface transmission electro-photonic device with a vertical cavity," IEEE J. Quantum Electron. **29**, 403 (1993)
60. T. Numai, K. Kurihara, I. Ogura, et al.: "Effect of sidewall reflector on current versus light-output in a pnpn vertical to surface transmission electro-photonic device with a vertical cavity," IEEE J. Quantum Electron. **29**, 2006 (1993).
61. T. Numai: "Analysis of photon recycling in semiconductor ring lasers," Jpn. J. Appl. Phys., Part 1, **39**, 6535 (2000)

For further reading, the following articles will be useful.

62. J. K. Butler, ed.: *Semiconductor Injection Lasers* (IEEE Press, New York 1979)
63. W. Streifer and M. Ettenberg, eds.: *Semiconductor Diode Lasers*, vol. 1 (IEEE Press, New York 1991)
64. H. Kressel and J. K. Butler: *Semiconductor Lasers and Heterojunction LEDs* (Academic Press, San Diego 1977)
65. H. C. Casey, Jr., and M. B. Panish: *Heterostructure Lasers A, B* (Academic Press, San Diego 1978)
66. G. H. B. Thompson: *Physics of Semiconductor Laser Devices* (John Wiley & Sons, New York 1980)
67. G. P. Agrawal and N. K. Dutta: *Long-Wavelength Semiconductor Lasers* (Van Nostrand Reinhold, New York 1986)
68. Y. Yamamoto, ed.: *Coherence, Amplification, and Quantum Effects in Semiconductor Lasers* (John Wiley & Sons, New York 1991)

69. G. P. Agrawal and N. K. Dutta: *Semiconductor Lasers*, 2nd edn (Van Nostrand Reinhold, New York 1993)

70. H. Kawaguchi: *Bistabilities and Nonlinearlities in Laser Diodes* (Artec House, Boston 1994)

71. W. W. Chow, S. W. Koch, and M. Sargent III: *Semiconductor-Laser Physics* (Springer, Berlin 1994)

72. G. P. Agrawal, ed.: *Semiconductor Lasers Past, Present, and Future* (AIP Press, Woodbury 1995)

73. L. A. Coldren and S. W. Corzine: *Diode Lasers and Photonic Integrated Circuits* (John Wiley & Sons, New York 1995)

74. W. W. Chow and S. W. Koch: *Semiconductor-Laser Fundamentals* (Springer, Berlin 1999)

75. S. E. Miller and A. G. Chynoweth, eds.: *Optical Fiber Telecommunications* (Academic Press, San Diego 1979)

76. S. E. Miller and I. P. Kaminow, eds.: *Optical Fiber Telecommunications II* (Academic Press, San Diego 1988)

77. I. P. Kaminow and T. L. Koch, eds.: *Optical Fiber Telecommunications IIIA* (Academic Press, San Diego 1997)

78. I. P. Kaminow and T. L. Koch, eds.: *Optical Fiber Telecommunications IIIB* (Academic Press, San Diego 1997)

79. C. Lin, ed.: *Optoelectronic Technology and Lightwave Communications Systems* (Van Nostrand Reinhold, New York 1989)

80. G. Keiser: *Optical Fiber Communications*, 2nd edn (McGraw-Hill, New York 1991)

81. G. P. Agrawal: *Fiber-Optic Communication Systems*, 2nd edn (John Wiley & Sons, New York 1997)

82. A. Yariv: *Introduction to Optical Electronics*, 2nd edn (Saunders, New York 1976)

83. A. Yariv: *Optical Electronics*, 3rd edn (Saunders, New York 1985), 4th edn (Saunders, New York 1991)

84. A. Yariv: *Optical Electronics in Modern Communications*, 5th edn (Oxford University Press, New York 1997)

85. B. E. A. Saleh and M. C. Teich: *Fundamentals of Photonics* (John Wiley & Sons, New York 1991)

Index

Springer Series in
OPTICAL SCIENCES